Annals of Mathematics Studies

Number 72

INTRODUCTION TO ALGEBRAIC *K*-THEORY

BY

JOHN MILNOR

PRINCETON UNIVERSITY PRESS

AND

UNIVERSITY OF TOKYO PRESS

PRINCETON, NEW JERSEY

1971

L.C. Card: 74-161197
I.S.B.N.: 0-691-08101-8

A.M.S. 1970: Primary, 16A54; Secondary, 10A15, 13D15, 18F25, 20G10

Published in Japan exclusively by
University of Tokyo Press;
in other parts of the world by
Princeton University Press

Printed in the United States of America

To Norman Steenrod

PREFACE AND GUIDE TO THE LITERATURE

The name "algebraic K-theory" describes a branch of algebra which centers about two functors K_0 and K_1, which assign to each associative ring Λ an abelian group $K_0\Lambda$ or $K_1\Lambda$ respectively. The theory has been developed by many authors, but the work of Hyman Bass has been particularly noteworthy, and Bass's book *Algebraic K-theory* (Benjamin, 1968), is the most important source of information. Here is a selected list of further references:

D. S. Rim, *Modules over finite groups,* Annals of Math. 69 (1959), 700-712.

R. Swan, *Projective modules over finite groups,* Bull. Amer. Math. Soc. 65 (1959), 365-367.

H. Bass, K-theory and stable algebra, Publ. Math. I.H.E.S. 22 (1964), 5-60.

H. Bass, A. Heller, and R. Swan, *The Whitehead group of a polynomial extension,* Publ. Math. I.H.E.S. 22 (1964), 61-79.

H. Bass, *The Dirichlet unit theorem, induced characters, and Whitehead groups of finite groups,* Topology 4 (1966), 391-410.

H. Bass (with A. Roy), *Lectures on topics in algebraic K-theory,* Tata Institute, Bombay 1967.

H. Bass and M. P. Murthy, *Grothendieck groups and Picard groups of abelian group rings,* Annals of Math. 86 (1967), 16-73.

R. Swan, *Algebraic K-theory,* Lecture Notes in Math. 76, Springer 1968.

R. Swan (with E. G. Evans), *K-theory of finite groups and orders,* Lecture Notes in Math. 149, Springer 1970.

L. N. Vaseršteĭn, *On the stabilization of the general linear group over a ring,* Mat. Sbornik 79 (121), 405-424 (1969). (Translation v. 8, 383-400 (A.M.S.).)

The main purpose of the present notes is to define and study an analogous functor K_2, also from associative rings to abelian groups. The definition is suggested by work of Robert Steinberg. This functor K_2 is related to K_1 and K_0 for example by means of an exact sequence

$$\begin{aligned} K_2\mathfrak{a} &\to K_2\Lambda \to K_2(\Lambda/\mathfrak{a}) \\ \to K_1\mathfrak{a} &\to K_1\Lambda \to K_1(\Lambda/\mathfrak{a}) \\ \to K_0\mathfrak{a} &\to K_0\Lambda \to K_0(\Lambda/\mathfrak{a}), \end{aligned}$$

associated with any two-sided ideal $\mathfrak{a}$ in the ring Λ; where $K_2\,\mathfrak{a}$, $K_1\,\mathfrak{a}$ and $K_0\mathfrak{a}$ are suitably defined relative groups. Here is a list of references for K_2:

R. Steinberg, *Générateurs, relations et revêtements de groupes algebriques,* Colloq. Théorie des groupes algebriques, Bruxelles 1962, 113-127.

R. Steinberg (with J. Faulkner and R. Wilson), *Lectures on Chevalley groups* (mimeographed), Yale 1967.

C. Moore, *Group extensions of* p-*adic and adelic linear groups,* Publ. Math. I.H.E.S. 35 (1969), 5-74.

H. Matsumoto, *Sur les sous-groupes arithmétiques des groupes semi-simples déployés,* Ann. Sci. Éc. Norm. Sup. 4e serie, 2 (1969), 1-62.

H. Bass, K_2 *and symbols,* pp. 1-11 of Lecture Notes in Math. 108, Springer 1969.

M. Kervaire, *Multiplicateurs de Schur et* K-*théorie,* pp. 212-225 of *Essays on Topology and Related Topics,* dedicated to G. de Rham (ed. A. Haefliger and R. Narasimhan), Springer 1970.

J. Wagoner, *On* K_2 *of the Laurent polynomial ring,* to appear.

B. J. Birch, K_2 *of global fields,* Proc. Symp. Pure Math. 20, Amer. Math. Soc. 1970.

J. Tate, *Symbols in arithmetic,* Proc. Int. Congr. Math. Nice, to appear.

M. Stein, *Chevalley groups over commutative rings.* Bull. Amer. Math. Soc. 77 (1971), 247-252.

It should be pointed out that definitions of K_n for all integers $n \geq 0$ have been proposed by several authors. See the following:

A. Nóbile and O. Villamayor, *Sur la* K-*théorie algébrique,* Ann. Sci. Éc. Norm. Sup. 4e série 1(1968), 581-616.

R. Swan, *Non-abelian homological algebra and* K-*theory,* Proc. Symp. in Pure Math. 17, 88-123, A.M.S. 1970.

M. Karoubi and O. Villamayor, *Foncteurs* K^n *en algèbre et en topologie,* C. R. Acad. Sc. Paris 269 (1969), 416-419.

S. Gersten, *Stable* K-*theory of discrete rings:* I and II, to appear.

J. Milnor, *Algebraic* K-*theory and quadratic forms,* Inventiones math. 9 (1970), 318-344.

D. Quillen, *The* K-*theory associated to a finite field:* I (mimeographed), 1970.

R. Swan, *Some relations between higher* K-*functors,* to appear.

These definitions are not mutually compatible, in general. Much work remains to be done in clarifying the relationships between various definitions. Note also that functors K_n for $n < 0$ have been defined by Bass (*Algebraic* K-*theory,* pp. 657-677).

The functors K_0 and K_1 are both important to geometric topologists. In the topological applications the ring Λ is always an integral group ring $Z\Pi$, where Π is the fundamental group of the object being studied. This theory had its beginnings in J.H.C. Whitehead's definition of the *torsion* associated with a homotopy equivalence between finite complexes. The Whitehead torsion lies in a certain factor group of $K_1 Z\Pi$. An important further step was taken by C. T. C. Wall. Consider a topological space A which is dominated by a finite complex. Then one can define a generalized "euler characteristic" $\chi(A)$, belonging to $K_0Z\Pi$. Wall showed that A has the homotopy type of a finite complex if and only if $\chi(A)$ is an integer. Siebenmann and Golo have shown that similar obstructions exist to the problem of fitting a boundary onto an open manifold.

Recent work by J. Wagoner and A. Hatcher indicates that the functor K_2 has similar topological applications. If one is given a "pseudo-isotopy" of a closed manifold, then an obstruction to deforming it into an isotopy lies in an appropriate factor group of $K_2Z\Pi$. Here is a list of references:

C. T. C. Wall, *Finiteness conditions for CW-complexes* I, Annals of Math. 81 (1965), 56-59; and II, Proc. Royal Soc. A 295 (1966), 129-139.

L. Siebenmann, *The structure of tame ends,* Notices Amer. Math. Soc. 13 (1966), 862.

J. Milnor, *Whitehead torsion,* Bull. Amer. Math. Soc. 72 (1966), 358-426.

G. de Rham, S. Maumary, and M. Kervaire, *Torsion et type simple d'homotopie,* Lecture Notes in Math. 48, Springer 1967.

V. L. Golo, *An invariant of open manifolds,* Izv. Akad. Nauk SSSR Ser. Mat. 31 (1967), 1091-1104. (Translation v. 1, 1041-1054 (A.M.S.).)

L. Siebenmann, *Torsion invariants for pseudo-isotopies on closed manifolds,* Notices Amer. Math. Soc. 14 (1967), 942.

R. M. F. Moss and C. B. Thomas (editors), *Algebraic* K-*theory and its geometric applications,* Lecture Notes in Math. 108, Springer 1969.

J. Wagoner, *Algebraic invariants for pseudo-isotopies, Proceed- of Liverpool Singularities Symposium II,* Lecture Notes in Math., Springer, to appear.

A strong impetus to the development of algebraic K-theory has been provided by work on the congruence subgroup problem, that is the problem of deciding whether every subgroup of finite index in an arithmetic group (such as SL(n, Λ) where Λ is the ring of integers in a number field) contains a congruence subgroup. This is closely related to the problem of computing $K_1\mathfrak{a}$ for an arbitrarily small ideal $\mathfrak{a} \subset \Lambda$. See the following, as well as the papers of Moore and Matsumoto mentioned earlier:

J. Mennicke, *Finite factor groups of the unimodular group,* Annals of Math. 81 (1965), 31-37.

J.-P. Serre, *Groupes de congruence,* Séminaire Bourbaki, 19e année (1966-67), n° 330.

H. Bass, *The congruence subgroup problem,* pp. 16-22 of *Local fields,* edited by T. A. Springer, Springer 1967.

H. Bass, J. Milnor, and J.-P. Serre, *Solution of the congruence subgroup problem for* SL_n $(n \geq 3)$ *and* Sp_{2n} $(n \geq 2)$, Publ. Math. I.H.E.S. 33 (1967).

L. N. Vaseršteĭn, K_1*-theory and the congruence subgroup problem,* Mat. Zametki 5 (1969), 233-244 (Russian).

J.-P. Serre, Le problème des groupes de congruence pour SL_2, Annals of Math. 92 (1970), 487-527.

I want to thank Hyman Bass, Robert Steinberg, and John Tate for many valuable conversations, and particularly for access to their unpublished work. Also I want to thank Jeffrey Joel for a number of suggestions, and for his lecture notes (based on lectures at Princeton University in 1967), which provided the starting point for this manuscript. Finally I want to thank Princeton University, U.C.L.A., M.I.T., and the Institute for Advanced Study, as well as the National Science Foundation (grants G.P. -7917, -13630, and -23305) for their support during the preparation of this manuscript.

CONTENTS

INTRODUCTION TO ALGEBRAIC *K*-THEORY

§1. Projective Modules and $K_0\Lambda$

The word *ring* will always mean associative ring with an identity element 1.

Consider left modules over a ring Λ. Recall that a module M is *free* if there exists a basis $\{m_\alpha\}$ so that each module element can be expressed uniquely as a finite sum $\Sigma\lambda_\alpha m_\alpha$, and *projective* if there exists a module N so that the direct sum $M \oplus N$ is free. This is equivalent to the requirement that every short exact sequence $0 \to X \to Y \to M \to 0$ must be split exact, so that $Y \cong X \oplus M$.

The *projective module group* $K_0\Lambda$ is an additive group defined by generators and relations as follows. There is to be one generator [P] for each isomorphism class of finitely generated projective modules P over Λ, and one relation

$$[P] + [Q] = [P \oplus Q]$$

for each pair of finitely generated projectives. (Compare the proof of 1.1 below.)

Clearly every element of $K_0\Lambda$ can be expressed as the difference $[P_1] - [P_2]$ of two generators. (In fact, adding the same projective module to P_1 and P_2 if necessary, we may even assume that P_2 is free.) We will give a criterion for the equality of two such differences.

First another definition. Let Λ^r denote the free module consisting of all r-tuples of elements of Λ. Two modules M and N are called *stably isomorphic* if there exists an integer r so that

$$M \oplus \Lambda^r \cong N \oplus \Lambda^r.$$

LEMMA 1.1. *The generator* $[P]$ *of* $K_0\Lambda$ *is equal to the generator* $[Q]$ *if and only if* P *is stably isomorphic to* Q. *Hence the difference* $[P_1] - [P_2]$ *is equal to* $[Q_1] - [Q_2]$ *if and only if* $P_1 \oplus Q_2$ *is stably isomorphic to* $P_2 \oplus Q_1$.

Proof. The group $K_0\Lambda$ can be defined more formally as a quotient group F/R, where F is free abelian with one generator $\langle P\rangle$ for each isomorphism class of finitely generated projectives P, and where R is the subgroup spanned by all $\langle P\rangle + \langle Q\rangle - \langle P \oplus Q\rangle$. (Thus we are reserving the symbol $[P]$ for the residue class of $\langle P\rangle$ modulo R.) Note that a sum $\langle P_1\rangle + \dots + \langle P_k\rangle$ in F is equal to $\langle Q_1\rangle + \dots + \langle Q_k\rangle$ if and only if

$$P_1 \cong Q_{\pi(1)}, \dots, P_k \cong Q_{\pi(k)}$$

for some permutation π of $\{1,\dots,k\}$. If this is the case, note the resulting isomorphism

$$P_1 \oplus \dots \oplus P_k \cong Q_1 \oplus \dots \oplus Q_k.$$

Now suppose that $\langle M\rangle \equiv \langle N\rangle \bmod R$. This means that

$$\begin{aligned}\langle M\rangle - \langle N\rangle = \ & \Sigma(\langle P_i\rangle + \langle Q_i\rangle - \langle P_i \oplus Q_i\rangle) \\ & - \Sigma(\langle P_j'\rangle + \langle Q_j'\rangle - \langle P_j' \oplus Q_j'\rangle)\end{aligned}$$

for appropriate modules P_i, Q_i, P_j', Q_j'.

Transposing all negative terms to the opposite side of the equation and then applying the remark above, we get

$$M \oplus (\Sigma (P_i \oplus Q_i) \oplus \Sigma P_j' \oplus \Sigma Q_j') \cong N \oplus (\Sigma P_i \oplus \Sigma Q_i \oplus \Sigma (P_j' \oplus Q_j')),$$

or briefly $M \oplus X \cong N \oplus X$, since the expressions inside the long parentheses are clearly isomorphic. Now choose Y so that $X \oplus Y$ is free, say $X \oplus Y \cong \Lambda^r$. Then adding Y to both sides we obtain $M \oplus \Lambda^r \cong N \oplus \Lambda^r$. Thus M is stably isomorphic to N.

The rest of the proof of 1.1 is straightforward. ∎

If the ring Λ is commutative, note that the tensor product over Λ of (finitely generated projective) left Λ-modules is again a (finitely generated projective) left Λ module. Defining

$$[P]\cdot[Q] = [P \otimes Q]$$

we make the additive group $K_0\Lambda$ into a commutative ring. The identity element of this ring is the class $[\Lambda^1]$ of the free module on one generator.

In order to compute the group $K_0\Lambda$ it is necessary to ask two questions.

Question 1. Is every finitely generated projective over Λ actually free (or at least stably free)?

Question 2. Is the number of elements in a basis for a free module actually an invariant of the module? In other words if $\Lambda^r \cong \Lambda^s$ does it follow that $r = s$?

If both questions have an affirmative answer then clearly $K_0\Lambda$ *is the free abelian group generated by* $[\Lambda^1]$. *This will be true, for example, if* Λ *is a field, or a skew field, or a principal ideal domain.*

Of course Questions 1 and 2 may have negative answers. For example if Λ is the ring of endomorphisms of a finite dimensional vector space of dimension greater than 1, then Question 1 has a negative answer; and if Λ is the ring of endomorphisms of an infinite dimensional vector space then Question 2 has a negative answer. (The group $K_0\Lambda$ is infinite cyclic but not generated by $[\Lambda^1]$ in the first case, and is zero in the second.)

Here is an important example in which $K_0\Lambda$ is free cyclic.

LEMMA 1.2. *If* Λ *is a local ring, then every finitely generated** *projective is free, and* $K_0\Lambda$ *is the free cyclic group generated by* $[\Lambda^1]$.

First recall the relevant definitions. A ring element u is called a *unit* if there exists a ring element v with $uv = vu = 1$. The set $\Lambda^\bullet$ consisting of all units in Λ evidently forms a multiplicative group.

Λ is called a *local ring* if the set $\mathfrak{m} = \Lambda - \Lambda^\bullet$ consisting of all non-units is a left ideal. It follows that $\mathfrak{m}$ is a right ideal also. For if some

---

* Compare Kaplansky, *Projective modules*, Annals of Mathematics 68 (1958), 372-377.

product $m\lambda$ with $m \in \mathfrak{m}$ and $\lambda \in \Lambda$ were a unit, then clearly m would have a right inverse, say $mv = 1$. This element v certainly cannot belong to the left ideal $\mathfrak{m}$. But v cannot be a unit either. For if v were a unit, then the computation

$$m = m(vv^{-1}) = (mv)v^{-1} = v^{-1}$$

would show that m must be a unit.

This contradiction shows that $\mathfrak{m}$ is indeed a two-sided ideal. The quotient ring $\Lambda/\mathfrak{m}$ is evidently a field or skew-field.

Note that a square matrix with entries in Λ is non-singular if and only if the corresponding matrix with entries in the quotient $\Lambda/\mathfrak{m}$ is non-singular. To prove this fact, multiply the given matrix on the left by a matrix which represents an inverse modulo $\mathfrak{m}$, and then apply elementary row operations to diagonalize. This shows that the matrix has a left inverse, and a similar argument constructs a right inverse.

We are now ready to prove Lemma 1.2. If the module P is finitely generated and projective over Λ then we can choose Q so that $P \oplus Q \cong \Lambda^r$. Thinking of the quotients $P/\mathfrak{m}P$ and $Q/\mathfrak{m}Q$ as vector spaces over the skew-field $\Lambda/\mathfrak{m}$, we can choose bases. Choose a representative in P or in Q for each basis element. The above remark on matrices then implies that the elements so obtained constitute a basis for $P \oplus Q$. Clearly it follows that P and Q are free. Since the dimension of the vector space $P/\mathfrak{m}P$ is an invariant of P, this completes the proof. ■

Next consider a homomorphism

$$f : \Lambda \to \Lambda'$$

between two rings. (It is always assumed that $f(1) = 1$.) Then every module M over Λ gives rise to a module

$$f_{\#}M = \Lambda' \otimes_{\Lambda} M$$

over Λ'. Clearly if M is finitely generated, or free, or projective, or splits as a direct sum over Λ, then $f_{\#}M$ is finitely generated, or free, or projective, or splits as as a corresponding direct sum over Λ'. Hence the correspondence

$$[P] \mapsto [f_{\#}P]$$

gives rise to a homomorphism

$$f_* : K_0\Lambda \to K_0\Lambda'$$

of abelian groups. Note the functorial properties

$$(\text{identity})_* = \text{identity}, \ (f \circ g)_* = f_* \circ g_*.$$

Example 1. Let Z be the ring of integers. Then for any ring Λ there is a unique homomorphism

$$i : Z \to \Lambda.$$

The image

$$i_* K_0 Z \subset K_0\Lambda$$

is clearly the subgroup generated by the free module $[\Lambda^1]$. The co-kernel

$$K_0\Lambda/(\text{subgroup generated by } [\Lambda^1]) = K_0\Lambda/i_* K_0 Z$$

is called the *projective class group* of Λ.

Example 2. Suppose that Λ can be mapped homomorphically into a field or skew-field F. This is always possible, for example, if Λ is commutative. Then we obtain a homomorphism

$$j_* : K_0\Lambda \to K_0 F \cong Z.$$

In the commutative case, this homomorphism is clearly determined by the kernel of j, which is a prime ideal in Λ. Hence one can speak of the *rank* of a projective module at a prime ideal $\mathfrak{p}$. If $\mathfrak{p} \supset \mathfrak{p}'$, note that the rank at $\mathfrak{p}$ is equal to the rank at $\mathfrak{p}'$. For if we localize the integral domain $\Lambda/\mathfrak{p}'$ at the ideal corresponding to $\mathfrak{p}$ (that is adjoin the inverses of all elements not belonging to $\mathfrak{p}$) we obtain a local ring which embeds in the quotient field of $\Lambda/\mathfrak{p}'$ and maps homomorphically into the quotient field of $\Lambda/\mathfrak{p}$. Using Lemma 1.2, it follows that the ranks are equal. *In particular, if* Λ *is an integral domain, then the rank of a projective module is the same at all prime ideals.*

In any case, choosing some fixed homomorphism $j : \Lambda \to F$, since $j_* i_*$ is an isomorphism, we obtain a direct sum decomposition

$$K_0\Lambda = (\text{image } i_*) \oplus (\text{kernel } j_*).$$

The first summand is free cyclic, and the second maps bijectively to the projective class group of Λ.

In the commutative case, note that (kernel j_*) is an ideal in the ring $K_0\Lambda$. We will denote this ideal by $\tilde{K}_0\Lambda$, and write

$$K_0\Lambda \cong Z \oplus \tilde{K}_0\Lambda.$$

Example 3. Suppose that Λ splits as a cartesian product

$$\Lambda_1 \times \Lambda_2 \times \dots \times \Lambda_k$$

of rings. Then the projection homomorphisms

$$K_0\Lambda \to K_0\Lambda_i$$

give rise to a corresponding cartesian product structure

$$K_0\Lambda \cong K_0\Lambda_1 \times K_0\Lambda_2 \times \dots \times K_0\Lambda_k.$$

The proof is not difficult.

Such a splitting of Λ occurs for example whenever Λ is commutative and artinian,* but is not local. For since Λ is commutative, the set of all nilpotent elements forms an ideal. If Λ is not local, there must exist an element λ which is neither a unit nor a nilpotent element. Since Λ is artinian, the sequence of principal ideals

$$(\lambda) \supset (\lambda^2) \supset (\lambda^3) \supset \dots$$

must terminate, say $(\lambda^n) = (\lambda^{n+1}) = \dots$ so that $\lambda^n = \rho\lambda^{2n}$ for some ρ. But this implies that the element $e = \rho\lambda^n$ is idempotent $(ee = e)$, and hence that Λ splits as a cartesian product

$$\Lambda \cong \Lambda/(e) \times \Lambda/(1 - e).$$

This splitting is not trivial since the hypothesis that λ is neither a unit nor nilpotent implies that $e \neq 1, 0$. This procedure can be continued inductively until Λ has been expressed as a cartesian product of local rings. It then follows that

$$K_0\Lambda \cong Z \times Z \times \dots \times Z.$$

---

* A ring is *artinian* if every descending sequence of ideals must terminate.

Dedekind Domains

Important examples in which the ring $K_0\Lambda$ has a more interesting structure are provided by Dedekind domains. We will discuss these in some detail, starting for variety with a non-standard version of the definition.*

DEFINITION. A *Dedekind domain* is a commutative ring without zero divisors such that, for any pair of ideals $\mathfrak{a} \subset \mathfrak{b}$, there exists an ideal $\mathfrak{c}$ with $\mathfrak{a} = \mathfrak{b}\mathfrak{c}$.

REMARK 1.3. Note that the ideal $\mathfrak{c}$ is uniquely determined, except in the trivial case $\mathfrak{a} = \mathfrak{b} = 0$. In fact if $\mathfrak{b}\,\mathfrak{c} = \mathfrak{b}\,\mathfrak{c}'$, then choosing some non-zero principal ideal $b_0\Lambda \subset \mathfrak{b}$ we can express $b_0\Lambda$ as a product $\mathfrak{x}\mathfrak{b}$ and conclude that $\mathfrak{x}\,\mathfrak{b}\mathfrak{c} = \mathfrak{x}\,\mathfrak{b}\,\mathfrak{c}'$, hence $b_0\mathfrak{c} = b_0\mathfrak{c}'$, from which the equality $\mathfrak{c} = \mathfrak{c}'$ follows.

DEFINITION. Two non-zero ideals $\mathfrak{a}$ and $\mathfrak{b}$ in the Dedekind domain Λ belong to the same *ideal class* if there exist non-zero ring elements x and y so that $x\mathfrak{a} = y\mathfrak{b}$.

Clearly the ideal classes of Λ form an abelian group under multiplication, with the class of principal ideals as identity element. We will use the notation $C(\Lambda)$ for the ideal class group of Λ, and the notation $\{\mathfrak{a}\} \in C(\Lambda)$ for the ideal class of $\mathfrak{a}$.

Note that $\{\mathfrak{a}\} = \{\mathfrak{b}\}$ if and only if $\mathfrak{a}$ is isomorphic, as Λ-module, to $\mathfrak{b}$. For if $\phi : \mathfrak{a} \to \mathfrak{b}$ is an isomorphism, then choosing $a_0 \in \mathfrak{a}$, the computation $a_0\phi(a) = \phi(a_0 a) = \phi(a_0)a$ shows that $a_0\mathfrak{b} = \phi(a_0)\mathfrak{a}$.

Important examples of Dedekind domains can be constructed as follows. Let F be a finite extension of the field Q of rational numbers. An element of F is called an *algebraic integer* if it is the root of a monic polynomial

---

* The usual definition is of course equivalent to the one given here. For further information, see Zariski and Samuel, *Commutative Algebra* I, Van Nostrand 1958; or Lang, *Algebraic Number Theory*, Addison-Wesley 1970; as well as Cartan and Eilenberg, *Homological Algebra*, Princeton University Press 1956.

$$x^k + a_1 x^{k-1} + \ldots + a_k$$

with coefficients $a_i \in Z$.

THEOREM 1.4. *The set* $\Lambda = \Lambda(F)$ *consisting of all algebraic integers in* F *is a Dedekind domain, with quotient field* F.

The proof of this classical theorem will be deferred until the end of §1.

For such a ring $\Lambda(F)$ of algebraic integers, the ideal class group $C(\Lambda(F))$ is always finite. (See for example Hecke, *Vorlesungen über die Theorie der algebraischen Zahlen*, Chelsea 1948; or Borevich and Shafarevich, *Number Theory*, Academic Press 1966.) As examples, for the domains $Z[i]$, $Z[\sqrt{-5}]$, and $Z[(1+\sqrt{-23})/2]$, the ideal class group has order 1, 2, and 3 respectively. Further examples will be given in §3.4.

Projective modules over Dedekind domains can be classified as follows.

LEMMA 1.5. *Every ideal in a Dedekind domain* Λ *is finitely generated and projective over* Λ. *Conversely every finitely generated projective module over* Λ *is isomorphic to a direct sum* $\mathfrak{a}_1 \oplus \ldots \oplus \mathfrak{a}_k$ *of ideals.*

Proof. If $\mathfrak{b}$ is a non-zero ideal, choose $0 \neq a_0 \in \mathfrak{b}$, so $a_0\Lambda \subset \mathfrak{b}$, and define $\mathfrak{c}$ by the equality $a_0\Lambda = \mathfrak{b}\,\mathfrak{c}$. Then the generator a_0 can be expressed as a finite sum $b_1c_1 + \ldots + b_kc_k$ with $b_i \in \mathfrak{b}$, $c_i \in \mathfrak{c}$. Define Λ-linear mappings

$$\mathfrak{b} \to \Lambda^k \text{ and } \Lambda^k \to \mathfrak{b}$$

by the formulas $b \mapsto (bc_1/a_0, \ldots, bc_k/a_0)$ and $(x_1, \ldots, x_k) \mapsto b_1x_1 + \ldots + b_kx_k$. Since the composition is the identity map of $\mathfrak{b}$, this proves that $\mathfrak{b}$ is finitely generated and projective.

Any finitely generated projective P can be embedded in the free module Λ^k for some k. Projecting to the k-th factor we obtain a homomorphism $\phi : P \to \Lambda$ with (kernel ϕ) $\subset \Lambda^{k-1}$.

Since the image $\phi(P) = \mathfrak{a}_k$ is an ideal, hence projective, we have $P \cong (\text{kernel } \phi) \oplus \mathfrak{a}_k$. An easy induction now completes the proof. ∎

REMARK. More generally, any module which is finitely generated and torsion free over Λ can easily be embedded in some Λ^k and hence, by this argument, is projective.

THEOREM 1.6. (Steinitz). *Two direct sums* $\mathfrak{a}_1 \oplus \dots \oplus \mathfrak{a}_r$ *and* $\mathfrak{b}_1 \oplus \dots \oplus \mathfrak{b}_s$ *of non-zero ideals are isomorphic as* Λ*-modules if and only if* $r = s$ *and the ideal class* $\{\mathfrak{a}_1 \mathfrak{a}_2 \dots \mathfrak{a}_r\}$ *is equal to* $\{\mathfrak{b}_1 \mathfrak{b}_2 \dots \mathfrak{b}_r\}$.

(Compare Kaplansky, *Modules over Dedekind rings and valuation rings*, Trans. Amer. Math. Soc. 72 (1952), 327-340.)

For the first half of the proof, the ring Λ can be any integral domain. First note that, if $\mathfrak{a} \subset \Lambda$ is a non-zero ideal, then any Λ-linear mapping $\phi : \mathfrak{a} \to \mathfrak{b} \subset \Lambda$ determines a unique element q of the quotient field of Λ such that

$$\phi(a) = qa \text{ for all } a \in \mathfrak{a}.$$

To prove this it is only necessary to divide the equation $a_0\phi(a) = \phi(a_0 a) = \phi(a_0)a$ by a_0, setting $q = \phi(a_0)/a_0$. Similarly, if the mapping

$$\phi : \mathfrak{a}_1 \oplus \dots \oplus \mathfrak{a}_r \to \mathfrak{b}_1 \oplus \dots \oplus \mathfrak{b}_s$$

is Λ-linear, then there is a unique $s \times r$ matrix $Q = (q_{ij})$ with entries in the quotient field so that the i-th component of $\phi(a_1, \dots, a_r) = (b_1, \dots, b_s)$ is

$$b_i = \Sigma\, q_{ij} a_j$$

for all $(a_1, \dots, a_r) \in \mathfrak{a}_1 \oplus \dots \oplus \mathfrak{a}_r$. If ϕ is an isomorphism, then this matrix Q has an inverse, hence $r = s$. We then assert that the product ideal $\mathfrak{b}_1 \dots \mathfrak{b}_r$ is equal to $(\det Q)\, \mathfrak{a}_1 \dots \mathfrak{a}_r$. In fact for each generator $a_1 \dots a_r$ of $\mathfrak{a}_1 \dots \mathfrak{a}_r$, the product $(\det Q)\, a_1 \dots a_r$ can be expressed as the determinant of the product matrix

$$Q\begin{pmatrix} a_1 0 \dots 0 \\ 0\ a_2 \cdots 0 \\ \cdot \quad \cdot \quad \cdot \\ \cdot \quad \cdot \quad \cdot \\ \cdot \quad \cdot \quad \cdot \\ 0\ 0 \dots a_r \end{pmatrix},$$

whose i-th row consists completely of elements $q_{ij}a_j$ of $\mathfrak{b}_i$. This proves that

$$(\det Q)\mathfrak{a}_1 \dots \mathfrak{a}_r \subset \mathfrak{b}_1 \dots \mathfrak{b}_r.$$

A similar argument shows that

$$(\det Q^{-1})\mathfrak{b}_1 \dots \mathfrak{b}_r \subset \mathfrak{a}_1 \dots \mathfrak{a}_r.$$

Multiplying this last inclusion by $\det Q$ and comparing, it follows that $\mathfrak{b}_1 \dots \mathfrak{b}_r$ is equal to $(\det Q)\mathfrak{a}_1 \dots \mathfrak{a}_r$; and hence belongs to the ideal class $\{\mathfrak{a}_1 \dots \mathfrak{a}_r\}$. This proves the first half of 1.6.

To prove that the rank r and the ideal class $\{\mathfrak{a}_1 \dots \mathfrak{a}_r\}$ form a complete invariant for $\mathfrak{a}_1 \oplus \dots \oplus \mathfrak{a}_r$, it clearly suffices to prove the following.

LEMMA 1.7. *If $\mathfrak{a}$ and $\mathfrak{b}$ are non-zero ideals in a Dedekind domain Λ, then the module $\mathfrak{a} \oplus \mathfrak{b}$ is isomorphic to $\Lambda^1 \oplus (\mathfrak{a}\mathfrak{b})$.*

If $\mathfrak{a}$ and $\mathfrak{b}$ happen to be relatively prime $(\mathfrak{a} + \mathfrak{b} = \Lambda)$, the proof proceeds as follows. Map $\mathfrak{a} \oplus \mathfrak{b}$ onto Λ^1 by the correspondence $a \oplus b \mapsto a + b$. The kernel is clearly isomorphic to the module $\mathfrak{a} \cap \mathfrak{b}$. Since Λ^1 is projective, the sequence $0 \to \mathfrak{a} \cap \mathfrak{b} \to \mathfrak{a} \oplus \mathfrak{b} \to \Lambda^1 \to 0$ is split exact, and therefore $\mathfrak{a} \oplus \mathfrak{b} \cong \Lambda^1 \oplus (\mathfrak{a} \cap \mathfrak{b})$. But the intersection $\mathfrak{a} \cap \mathfrak{b}$ is equal to the product ideal $\mathfrak{a}\mathfrak{b}$. For the inclusion $\mathfrak{a}\mathfrak{b} \subset \mathfrak{a} \cap \mathfrak{b}$ is clear; and if $1 = a_0 + b_0$ then every $x \in \mathfrak{a} \cap \mathfrak{b}$ can be expressed as $x = a_0x + xb_0$, and hence belongs to $\mathfrak{a}\mathfrak{b}$. Thus $\mathfrak{a} \oplus \mathfrak{b} \cong \Lambda^1 \oplus \mathfrak{a}\mathfrak{b}$ as required.

For the general case, the hypothesis that Λ is a Dedekind domain will be needed in order to replace $\mathfrak{a}$ by an ideal which is relatively prime to $\mathfrak{b}$. It clearly suffices to prove the following.

LEMMA 1.8. *Given non-zero ideals $\mathfrak{a}$ and $\mathfrak{b}$ in a Dedekind domain Λ there exists an ideal $\mathfrak{a}'$ in the ideal class of $\mathfrak{a}$ which is prime to $\mathfrak{b}$.*

To prove this we must first establish two of the standard properties of Dedekind domains.

LEMMA 1.9. *Every non-zero ideal in a Dedekind domain* Λ *can be expressed uniquely as a product of maximal ideals.*

In fact choosing any maximal ideal $\mathfrak{m}_1 \supset \mathfrak{a}$ we have $\mathfrak{a} = \mathfrak{m}_1\mathfrak{a}_1$ for some ideal $\mathfrak{a}_1$, then similarly $\mathfrak{a}_1 = \mathfrak{m}_2\mathfrak{a}_2$, and so on, with $\mathfrak{a} \subset \mathfrak{a}_1 \subset \mathfrak{a}_2 \subset \ldots$ This sequence must terminate, since Λ is Noetherian by 1.5. The resulting factorization is unique. For if $\mathfrak{m}_1 \ldots \mathfrak{m}_k = \mathfrak{m}'_1 \ldots \mathfrak{m}'_\ell$ then $\mathfrak{m}'_1 \supset \mathfrak{m}_1 \ldots \mathfrak{m}_k$ and hence, since $\mathfrak{m}'_1$ is prime, $\mathfrak{m}'_1$ contains some $\mathfrak{m}_i$, and therefore is equal to $\mathfrak{m}_i$. The uniqueness statement then follows by induction on $\mathrm{Max}(k,\ell)$, using 1.3. ∎

LEMMA 1.10. *For any non-zero ideal* $\mathfrak{a}$ *in a Dedekind domain* Λ, *the quotient* $\Lambda/\mathfrak{a}$ *is a principal ideal ring (usually with zero divisors).*

Proof. Let $\mathfrak{m}_1, \ldots, \mathfrak{m}_k$ be the distinct maximal ideals containing $\mathfrak{a}$. We will first show that each $\mathfrak{m}_i$ is a principal ideal modulo $\mathfrak{a}$. Let x_1 be a ring element which belongs to $\mathfrak{m}_1$ but not to ${\mathfrak{m}_1}^2$. Since the ideals ${\mathfrak{m}_1}^2, \mathfrak{m}_2, \ldots, \mathfrak{m}_k$ are pairwise relatively prime (using 1.9), it follows that there exists a ring element y_1 so that

$$y_1 \equiv x_1 \mod {\mathfrak{m}_1}^2,$$

$$y_1 \equiv 1 \mod \mathfrak{m}_j \text{ for } j > 1,$$

using the Chinese Remainder Theorem. (See for example Lang, *Algebra*, Addison-Wesley 1965.) Then the ideal generated by y_1 and $\mathfrak{a}$ is contained in $\mathfrak{m}_1$, but is not contained in ${\mathfrak{m}_1}^2$ or in any other maximal ideal. So, using 1.9., this ideal can only be $\mathfrak{m}_1$ itself.

This proves that $\mathfrak{m}_1$ is a principal ideal modulo $\mathfrak{a}$. But every ideal of $\Lambda/\mathfrak{a}$ is a product of maximal ideals, so this completes the proof of 1.10. ∎

We are now ready to prove Lemma 1.8. Given non-zero ideals $\mathfrak{a}$ and $\mathfrak{b}$, choose $0 \neq a_0 \in \mathfrak{a}$ and define $\mathfrak{x}$ by the equation $\mathfrak{x}\mathfrak{a} = a_0\Lambda$. Applying

1.10 to the ideal $\mathfrak{x}$ modulo $\mathfrak{b}\mathfrak{x}$, we see that $\mathfrak{x}$ is generated by $\mathfrak{b}\mathfrak{x}$ together with some element x_0. Now multiplying the equation

$$\mathfrak{x} = \mathfrak{b}\mathfrak{x} + x_0\Lambda$$

by $\mathfrak{a}$, and then dividing by a_0, we obtain

$$\Lambda = \mathfrak{b} + \mathfrak{a}x_0/a_0.$$

Since $\mathfrak{a}x_0/a_0$ is clearly an ideal in the ideal class $\{\mathfrak{a}\}$, this proves 1.8, and completes the proof of Theorem 1.6. ∎

COROLLARY 1.11. *If* Λ *is a Dedekind domain, then* $K_0\Lambda \cong Z \oplus \tilde{K}_0\Lambda$, *where the additive group of* $\tilde{K}_0\Lambda$ *is canonically isomorphic to the ideal class group* $C(\Lambda)$, *and where the product of any two elements in the ideal* $\tilde{K}_0\Lambda$ *is zero.*

In fact the correspondence

$$[\mathfrak{a}_1 \oplus \dots \oplus \mathfrak{a}_r] \mapsto (r, \{\mathfrak{a}_1 \dots \mathfrak{a}_r\})$$

maps $K_0\Lambda$ isomorphically onto $Z \oplus C(\Lambda)$. Recall that $\tilde{K}_0\Lambda$ can be identified with the set of differences $[P] - [Q]$ with rank $P =$ rank Q. Then each element of $\tilde{K}_0\Lambda$ can be written as a difference $[\mathfrak{a}] - [\Lambda^1]$, and we must prove that

$$([\mathfrak{a}] - [\Lambda^1])([\mathfrak{b}] - [\Lambda^1]) = 0.$$

But the product $[\mathfrak{a}][\mathfrak{b}] = [\mathfrak{a} \otimes \mathfrak{b}]$ is equal to $[\mathfrak{a}\mathfrak{b}]$. In fact the projective modules $\mathfrak{a} \otimes \mathfrak{b}$ and $\mathfrak{a}\mathfrak{b}$ both have rank 1, so the natural surjection from the tensor product to the product ideal is an isomorphism. The conclusion now follows from 1.7. ∎

Remarks. The ideal class group $C(\Lambda)$ can be naturally identified with a multiplicative group, $1 + \tilde{K}_0\Lambda$, of units in the ring $K_0\Lambda$. Something similar happens for an arbitrary commutative ring. Call a module M, over a commutative ring Λ, *invertible* if there exists a module N so that $M \otimes N$ is free on one generator. The set of isomorphism classes of invertible modules clearly forms a group under the tensor product. This group is called the *Picard group*, denoted by $\mathrm{Pic}(\Lambda)$.

It can be shown that a module is invertible if and only if it is projective, finitely generated, and has rank 1 at every prime ideal. (Compare Bourbaki, XXVII *Algèbre commutative*, Ch. 2, p. 143.) Furthermore the second exterior power $E^2_\Lambda M$ of an invertible module is zero. For this exterior power is a projective module which has rank zero at every prime. (*Algèbre commutative*, Ch. 2, p. 112.) It follows that the Picard group embeds as a subgroup of the group of units in $K_0\Lambda$. For if two invertible modules M and M' are stably isomorphic, $M \oplus \Lambda^r \cong M' \oplus \Lambda^r$, then taking the $(r+1)$-st exterior power of each side, we see that $M \cong M'$. (Bass proves the sharper statement that there exists a canonical retracting homomorphism from the additive group of $K_0\Lambda$ to the multiplicative group $\operatorname{Pic}(\Lambda) \subset K_0\Lambda$.)

In the case of a Dedekind domain, it is clear that $\operatorname{Pic}(\Lambda)$ is canonically isomorphic to the ideal class group $C(\Lambda)$.

To conclude §1, let us prove Theorem 1.4. If F is a finite extension of the field of rational numbers, we must show that the set Λ, consisting of all algebraic integers in F, is a Dedekind domain.

Let n be the degree of F over Q. It will be convenient to use the word *lattice* to mean an additive subgroup of F which has a finite basis. Thus every lattice L in F is a free abelian additive group of rank $\leq n$. The *product* LL' of two lattices in F is the lattice generated by all products $\ell\ell'$ with $\ell \in L$ and $\ell' \in L'$.

LEMMA 1.12. *An element* f *of* F *is an algebraic integer if and only if there exists a non-zero lattice* $L \subset F$ *with* $fL \subset L$.

For if f is a root of the polynomial $x^k + a_1x^{k-1} + \ldots + a_k$ with coefficients in Z, then the field elements $1, f, f^2, \ldots f^{k-1}$ span a lattice $L = Z[f]$ with $fL \subset L$. Conversely, if $fL \subset L$ where L is spanned by $b_1, \ldots, b_k$, then we can set

$$fb_i = \sum_j a_{ij}\, b_j$$

for some matrix (a_{ij}) of rational integers. Writing this as

$$\sum_j (f\delta_{ij} - a_{ij})\, b_j = 0 ,$$

where (δ_{ij}) denotes the $k \times k$ identity matrix, it follows that the columns of the matrix $(f\delta_{ij} - a_{ij})$ are linearly dependent. Therefore f satisfies the monic polynomial equation

$$\det(f\delta_{ij} - a_{ij}) = 0,$$

with coefficients in Z; which proves 1.12. ∎

It follows that the set Λ, consisting of all algebraic integers in F, is closed under addition and multiplication. For if $\lambda, \mu \in \Lambda$, then there exist lattices L and L′ with

$$\lambda L \subset L, \quad \mu L' \subset L'.$$

Now the product lattice $L'' = LL'$ will satisfy $(\lambda+\mu)L'' \subset L''$ and $\lambda\mu L'' \subset L''$. Thus Λ is a ring.

LEMMA 1.13. *This set $\Lambda \subset F$ is itself a lattice of rank n in F.*

The proof will be based on the trace homomorphism from F to Q. If $F' \supset F$ is a Galois extension of Q, recall that the additive homomorphism

$$\mathrm{trace}_{F/Q} : F \to Q$$

can be defined by the formula

$$\mathrm{trace}_{F/Q}(f) = \sigma_1(f) + \ldots + \sigma_n(f),$$

where $\sigma_1, \ldots, \sigma_n$ are the distinct embeddings of F in F′. (See, for example, Lang, *Algebra*.)

Note that the set of algebraic integers in Q is precisely equal to Z. For if a fraction a/b satisfies a monic polynomial equation with coefficients in Z, then clearing denominators we see that every prime which divides b must also divide a.

Therefore the trace homomorphism from F to Q maps Λ into Z. For if $\lambda \in \Lambda$, then $\mathrm{trace}_{F/Q}(\lambda)$ is both an algebraic integer (in F′) and a rational number; hence $\mathrm{trace}_{F/Q}(\lambda) \in Z$.

Note that every field element f possesses a multiple $f + f + \ldots + f = mf$ which is an algebraic integer. In fact, expressing f as the root of a polynomial with integer coefficients, we can take m to be the absolute value of the leading coefficient. It follows that the quotient field of Λ is equal to F.

Consider the Q-bilinear pairing

$$f, f' \mapsto \text{trace}_{F/Q}(ff')$$

from $F \times F$ to Q. This pairing is non-degenerate, since for each $f \neq 0$ we can choose $f' = 1/f$ so that $\text{trace}(ff') \neq 0$. Choose algebraic integers $\lambda_1, \ldots, \lambda_n$ which form a basis for F over Q. Then the Q-linear function

$$f \mapsto (\text{trace}_{F/Q}(\lambda_1 f), \ldots, \text{trace}_{F/Q}(\lambda_n f))$$

from F to $Q \oplus \ldots \oplus Q$ is bijective, and embeds Λ in the direct sum $Z \oplus \ldots \oplus Z$. Therefore Λ is finitely generated as additive group; which proves Lemma 1.13. ∎

It follows that every non-zero ideal $\mathfrak{a} \subset \Lambda$ is also a lattice of rank n in F. Here are three important consequences of 1.13.

(1) *The ring* Λ *is Noetherian.*

In fact, if $\mathfrak{a}$ is a non-zero ideal, then $\Lambda/\mathfrak{a}$ is finite, so there are only finitely many larger ideals.

(2) *Every non-zero prime ideal of* Λ *is maximal.*

For the quotient ring $\Lambda/\mathfrak{p}$, being finite and without zero divisors, must be a field.

(3) *If an element* f *in the quotient field of* Λ *satisfies* $f\mathfrak{a} \subset \mathfrak{a}$ *for some non-zero ideal* $\mathfrak{a}$, *then* $f \in \Lambda$.

In other words Λ is integrally closed in its quotient field. This follows using 1.12.

We will show that any domain satisfying (1), (2) *and* (3) *is necessarily Dedekind.* The proof, due to van der Waerden, is based on the following.

OBSERVATION. *Every non-zero ideal* $\mathfrak{a}$ *in a commutative Noetherian ring contains a product of non-zero prime ideals.*

For if $\mathfrak{a}$ itself is prime, there is nothing to prove. Otherwise, choosing ring elements λ and μ not in $\mathfrak{a}$ so that $\lambda\mu \in \mathfrak{a}$, the two ideals $\mathfrak{a} + \lambda\Lambda$ and $\mathfrak{a} + \mu\Lambda$ are strictly larger than $\mathfrak{a}$, but have product contained in $\mathfrak{a}$. Assuming inductively that the Observation is true for these two larger ideals, it follows that it is true for a also. (This "induction" argument makes sense since Λ is Noetherian.)

Now given a domain Λ satisfying (1), (2) and (3), and given non-zero ideals $\mathfrak{a} \subset \mathfrak{b}$, we must show that $\mathfrak{a} = \mathfrak{b}\mathfrak{c}$ for some ideal $\mathfrak{c}$. We will assume inductively that this statement is true for any ideal $\mathfrak{b}'$ which is strictly larger than $\mathfrak{b}$; and for any $\mathfrak{a}' \subset \mathfrak{b}'$. To start the induction, the statement is certainly true when $\mathfrak{b} = \Lambda$.

Choose an element $b \neq 0$ in $\mathfrak{b}$, and choose a product $\mathfrak{p}_1 \dots \mathfrak{p}_r$ of maximal ideals so that $\mathfrak{p}_1 \dots \mathfrak{p}_r \subset \Lambda b$, with r minimal. Also choose a maximal ideal $\mathfrak{p} \supset \mathfrak{b}$. Then $\mathfrak{p}$ contains the product $\mathfrak{p}_1 \dots \mathfrak{p}_r$, hence $\mathfrak{p}$ contains some $\mathfrak{p}_i$, and therefore $\mathfrak{p} = \mathfrak{p}_i$. To fix our ideas, assume that $\mathfrak{p} = \mathfrak{p}_1$. The product $\mathfrak{p}_2 \dots \mathfrak{p}_r$ is not contained in Λb, since r is minimal, so there exists an element $c \in \mathfrak{p}_2 \dots \mathfrak{p}_r$ with $c \notin \Lambda b$. Evidently

$$c\mathfrak{b} \subset c\mathfrak{p} \subset \mathfrak{p}_2 \dots \mathfrak{p}_r\mathfrak{p} = \mathfrak{p}_1 \dots \mathfrak{p}_r \subset \Lambda b.$$

Therefore

$$(c/b)\mathfrak{b} \subset \Lambda$$

even though the element $c/b \in F$ does not belong to Λ. Consider the ideal

$$\mathfrak{b}' = b^{-1}(\Lambda b + \Lambda c)\mathfrak{b} = \mathfrak{b} + (c/b)\mathfrak{b}$$

in Λ. Since $c/b \notin \Lambda$, it follows from (3) that $\mathfrak{b}'$ is strictly larger than $\mathfrak{b}$. Therefore, by the induction hypothesis, given any $\mathfrak{a} \subset \mathfrak{b}$ the equation

$$\mathfrak{a} = \mathfrak{b}'\mathfrak{c}'$$

has a solution $\mathfrak{c}'$. Setting

$$\mathfrak{c} = b^{-1}(\Lambda b + \Lambda c)\mathfrak{c}' \subset F,$$

we have

$$\mathfrak{b}\mathfrak{c} = b^{-1}(\Lambda b + \Lambda c)\mathfrak{b}\mathfrak{c}' = \mathfrak{b}'\mathfrak{c}' = \mathfrak{a},$$

are required. This set $\mathfrak{c}$ is actually contained in Λ, since it satisfies the condition $\mathfrak{b}\mathfrak{c} \subset \mathfrak{b}$. (Compare (3).) This shows that Λ is a Dedekind domain, and completes the proof of Theorem 1.4. ■

§2. Constructing Projective Modules

Consider a commutative square of ring homomorphisms

$$\begin{array}{ccc} \Lambda & \xrightarrow{i_1} & \Lambda_1 \\ {\scriptstyle i_2}\downarrow & & \downarrow{\scriptstyle j_1} \\ \Lambda_2 & \xrightarrow{j_2} & \Lambda' \end{array}$$

satisfying the following two conditions.

HYPOTHESIS 1. Λ is the product of Λ_1 and Λ_2 over Λ'. In other words, given elements $\lambda_1 \in \Lambda_1$ and $\lambda_2 \in \Lambda_2$ with a common image $j_1(\lambda_1) = j_2(\lambda_2)$ in Λ', there is one and only one element $\lambda \in \Lambda$ which satisfies $i_1(\lambda) = \lambda_1$ and $i_2(\lambda) = \lambda_2$.

HYPOTHESIS 2. At least one of the two homomorphisms j_1 and j_2 is surjective.

Then we will show how to construct projective modules over Λ, using projective modules over Λ_1 and Λ_2 as building blocks.

NOTATIONS. If $f : \Lambda \to \Lambda_1$ is a ring homomorphism, and M is a left Λ-module, recall that the induced left Λ_1-module $\Lambda_1 \otimes_\Lambda M$ is denoted by $f_\# M$. There is a canonical Λ-linear map

$$f_* : M \to f_\# M$$

defined by

$$f_*(m) = 1 \otimes_\Lambda m.$$

As an example, if M is free over Λ, with basis $\{b_\alpha\}$, note that $f_\# M$ is free over Λ_1 with basis $\{f_* b_\alpha\}$.

Now consider a commutative square of rings satisfying Hypotheses 1 and 2.

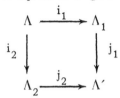

Basic construction. Given a projective module P_1 over Λ_1, a projective module P_2 over Λ_2, and given an isomorphism

$$h : j_{1\#} P_1 \to j_{2\#} P_2$$

over Λ', let $M = M(P_1, P_2, h)$ denote the subgroup of $P_1 \times P_2$ consisting of all pairs (p_1, p_2) with $hj_{1*}(p_1) = j_{2*}(p_2)$.

Thus we have a commutative square of additive groups

$$\begin{array}{ccc} M & \longrightarrow & P_1 \\ \downarrow & & \downarrow{\scriptstyle hj_{1*}} \\ P_2 & \xrightarrow{j_{2*}} & j_{2\#} P_2 \end{array}$$

satisfying the analogues of Hypotheses 1 and 2. We make M into a left Λ-module by setting

$$\lambda \cdot (p_1, p_2) = (i_1(\lambda) \cdot p_1, i_2(\lambda) \cdot p_2).$$

THEOREM 2.1. *The module* $M = M(P_1, P_2, h)$ *is projective over* Λ. *Furthermore if* P_1 *and* P_2 *are finitely generated over* Λ_1 *and* Λ_2 *respectively, then* M *is finitely generated over* Λ.

THEOREM 2.2. *Every projective* Λ*-module is isomorphic to* $M(P_1, P_2, h)$ *for some suitably chosen* P_1, P_2 *and* h.

THEOREM 2.3. *The modules* P_1 *and* P_2 *are naturally isomorphic to* $i_{1\#}M$ *and* $i_{2\#}M$ *respectively.*

The proofs will occupy the rest of §2.

Before proving Theorem 2.1 in general, we must first consider the special case where P_1 and P_2 are free modules.

Choose a basis $\{x_\alpha\}$ for P_1 over Λ_1 and a basis $\{y_\beta\}$ for P_2 over Λ_2. These determine corresponding bases $\{j_{1*} x_\alpha\}$ for $j_{1\#} P_1$ and $\{j_{2*} y_\beta\}$ for $j_{2\#} P_2$ over Λ'. The isomorphism

$$h : j_{1\#} P_1 \to j_{2\#} P_2$$

is now completely described by the matrix*

$$A = (a_{\alpha\beta})$$

over Λ', where

$$h(j_{1*}\, x_\alpha) = \Sigma\, a_{\alpha\beta}\, j_{2*}\, y_\beta.$$

A given matrix $(a_{\alpha\beta})$ can occur in this construction if and only if it is invertible (i.e., has a two sided inverse matrix $(b_{\beta\alpha})$).

LEMMA 2.4. *If this matrix* $A = (a_{\alpha\beta})$ *is the image under* j_2 *of an invertible matrix over* Λ_2, *then the module* $M = M(P_1,P_2,h)$ *is free.*

Proof. Let $a_{\alpha\beta} = j_2\, c_{\alpha\beta}$ where $(c_{\alpha\beta})$ is invertible. Set

$$y'_\alpha = \Sigma\, c_{\alpha\beta}\, y_\beta \in P_2.$$

Clearly these elements $\{y'_\alpha\}$ form a basis for P_2. The identity

$$hj_{1*}\, x_\alpha = j_{2*}\, y'_\alpha$$

now shows that the pair

$$z_\alpha = (x_\alpha, y'_\alpha) \in P_1 \times P_2$$

belongs to the submodule $M(P_1,P_2,h) \subset P_1 \times P_2$. It is now easy to verify that $M(P_1,P_2,h)$ is free over Λ with basis $\{z_\alpha\}$. ∎

LEMMA 2.5. *If* P_1 *and* P_2 *are free, and* j_2 *is surjective, then* $M(P_1,P_2,h)$ *is projective.*

Proof. Let Q_1 be free over Λ_1 with one basis element u_β for each element y_β of the basis of P_2. Similarly let Q_2 be free over Λ_2 with basis $\{v_\alpha\}$, corresponding to the basis $\{x_\alpha\}$ for P_1. Let

$$g : j_{1\#}\, Q_1 \to j_{2\#}\, Q_2$$

* We will be mainly interested in the case where the index sets $\{\alpha\}$ and $\{\beta\}$ are finite. However the argument works just as well for infinite index sets, providing that we make the convention that all infinite "matrices" are to have only finitely many non-zero entries in each row.

be the isomorphism with matrix A^{-1} where $A = (a_{\alpha\beta})$ is the matrix of h. Then

$$M(P_1,P_2,h) \oplus M(Q_1,Q_2,g) \cong M(P_1 \oplus Q_1,P_2 \oplus Q_2,h \oplus g),$$

where $h \oplus g$ corresponds to a matrix of the form

$$\begin{pmatrix} A & 0 \\ 0 & A^{-1} \end{pmatrix}$$

over Λ'. We will prove that this compound matrix satisfies the hypothesis of Lemma 2.4. Hence $M(P_1 \oplus Q_1,P_2 \oplus Q_2,h \oplus g)$ is free, and therefore $M(P_1,P_2,h)$ is projective.

Start with the identity

$$\begin{pmatrix} A & 0 \\ 0 & A^{-1} \end{pmatrix} = \begin{pmatrix} I & A \\ 0 & I \end{pmatrix}\begin{pmatrix} I & 0 \\ -A^{-1} & I \end{pmatrix}\begin{pmatrix} I & A \\ 0 & I \end{pmatrix}\begin{pmatrix} 0 & -I \\ I & 0 \end{pmatrix}.$$

Since j_2 is surjective, the first factor on the right can clearly be lifted to some matrix of the form $\begin{pmatrix} I & * \\ 0 & I \end{pmatrix}$ over Λ_2. But any matrix of this form is invertible. Since the other factors lift similarly, this proves 2.5. ∎

Now consider the general case of Theorem 2.1, where the modules P_1 and P_2 are only assumed to be projective, with

$$h : j_{1\#} P_1 \overset{\cong}{\to} j_{2\#} P_2.$$

LEMMA 2.6. *There exist projectives Q_1 over Λ_1 and Q_2 over Λ_2 so that $P_1 \oplus Q_1$ and $P_2 \oplus Q_2$ are free, and so that $j_{1\#} Q_1 \cong j_{2\#} Q_2$.*

Proof. Since P_1 is projective we can certainly choose some module N_1 over Λ_1 so that $P_1 \oplus N_1$ is free, say of rank r, over Λ_1. Here r can be any cardinal number.* We will write

$$P_1 \oplus N_1 \cong (\Lambda_1)^r.$$

* Note however that if P_1 is finitely generated then N_1 can be chosen so that r is finite.

Similarly we can choose N_2 so that

$$P_2 \oplus N_2 \cong (\Lambda_2)^s.$$

Now applying $j_{1\#}$ and $j_{2\#}$ respectively, and setting

$$P' = j_{1\#} P_1 \cong j_{2\#} P_2$$

we have

$$P' \oplus j_{1\#} N_1 \cong (\Lambda')^r$$

$$P' \oplus j_{2\#} N_2 \cong (\Lambda')^s$$

and therefore

$$j_{1\#} N_1 \oplus (\Lambda')^s \cong j_{1\#} N_1 \oplus P' \oplus j_{2\#} N_2 \cong j_{2\#} N_2 \oplus (\Lambda')^r.$$

Now defining

$$Q_1 = N_1 \oplus (\Lambda_1)^s, \; Q_2 = N_2 \oplus (\Lambda_2)^r$$

it follows that $j_{1\#} Q_1 \cong j_{2\#} Q_2$ as required. ■

Proof of Theorem 2.1. Choose Q_1 and Q_2 as above and choose some isomorphism

$$k : j_{1\#} Q_1 \to j_{2\#} Q_2.$$

Then, since j_2 is surjective, the module

$$M(P_1,P_2,h) \oplus M(Q_1,Q_2,k) \cong M(P_1 \oplus Q_1,P_2 \oplus Q_2,h \oplus k)$$

is projective by Lemma 2.5. Hence $M(P_1,P_2,h)$ must itself be projective.

If P_2 and P_2 are finitely generated, we must prove that $M(P_1,P_2,h)$ is also finitely generated. But we can certainly choose Q_1 and Q_2 to be finitely generated also. Looking through the proofs of Lemmas 2.4 and 2.5, we see that all of the constructions used preserve finite generation. Hence $M(P_1,P_2,h)$ is a direct summand of a finitely generated free module, and therefore is finitely generated. This completes the proof of Theorem 2.1. ■

Proof of Theorem 2.2. If P is projective over Λ set

$$P_1 = i_{1\#} P, \quad P_2 = i_{2\#} P.$$

Since $j_1 i_1 = j_2 i_2$ there is a canonical isomorphism

$$h : j_{1\#} P_1 \to j_{2\#} P_2.$$

The diagram

$$\begin{array}{ccc} P & \xrightarrow{i_{1*}} & P_1 \\ {\scriptstyle i_{2*}}\downarrow & & \downarrow{\scriptstyle hj_{1*}} \\ P_2 & \xrightarrow{j_{2*}} & j_{2\#} P_2 \end{array}$$

now clearly satisfies the analogues of Hypotheses 1 and 2. In particular, P is the product of P_1 and P_2 over $j_{2\#} P_2$. Hence

$$P \cong M(P_1, P_2, h). \blacksquare$$

Proof of Theorem 2.3. For any $M = M(P_1, P_2, h)$ the natural Λ-linear map $M \to P_1$ gives rise to a Λ_1-linear map

$$f : i_{1\#} M \to P_1.$$

We must prove that f is an isomorphism.

Special case. Under the hypotheses of Lemma 2.4, the modules $i_{1\#}M$ and P_1 are free over Λ_1 with bases which correspond under f. So certainly f is an isomorphism.

General case. The proof of Theorem 2.1 consisted in showing that every $M(P_1, P_2, h)$ is a direct summand of some $M(P_1 \oplus \bar{P}_1, P_2 \oplus \bar{P}_2, h \oplus \bar{h})$ which satisfies the hypotheses of Lemma 2.4. It follows that the map $f \oplus \bar{f}$ associated with this direct sum is an isomorphism; hence f itself must be an isomorphism. This completes the proof. $\blacksquare$

§3. The Whitehead Group $K_1\Lambda$

Let $GL(n,\Lambda)$ denote the general linear group consisting of all $n \times n$ invertible matrices over Λ, and let $GL(\Lambda)$ denote the direct limit (or union) of the sequence

$$GL(1,\Lambda) \subset GL(2,\Lambda) \subset GL(3,\Lambda) \subset \ldots,$$

where each $GL(n,\Lambda)$ is injected into $GL(n+1,\Lambda)$ by the correspondence

$$A \mapsto \begin{pmatrix} A & 0 \\ 0 & 1 \end{pmatrix}.$$

A matrix in $GL(\Lambda)$ is called *elementary* if it coincides with the identity matrix except for a single off-diagonal entry.

WHITEHEAD LEMMA 3.1. *The subgroup* $E(\Lambda) \subset GL(\Lambda)$ *generated by all elementary matrices is precisely equal to the commutator subgroup of* $GL(\Lambda)$.

Proof. It is easily verified that each elementary matrix can be expressed as a commutator of two other elementary matrices. (Compare §5.) Conversely every commutator $ABA^{-1}B^{-1}$ in $GL(n,\Lambda)$ can be expressed as a product

$$\begin{pmatrix} A & 0 \\ 0 & A^{-1} \end{pmatrix}\begin{pmatrix} B & 0 \\ 0 & B^{-1} \end{pmatrix}\begin{pmatrix} (BA)^{-1} & 0 \\ 0 & BA \end{pmatrix}$$

in $GL(2n,\Lambda)$, and it follows from the proof of 2.5 that each of these factors can be expressed as a product of elementary matrices. (Compare §4.3.) ∎

Hence $E(\Lambda)$ is a normal subgroup, and the quotient $GL(\Lambda)/E(\Lambda)$ is a well defined abelian group.

DEFINITION. This abelian quotient group $GL(\Lambda)/E(\Lambda)$ is called the *Whitehead group* $K_1\Lambda$.

We will sometimes think of $K_1\Lambda$ as an additive group, and sometimes as a multiplicative group.

Clearly any ring homomorphism $f : \Lambda \to \Lambda'$ gives rise to a group homomorphism $f_* : K_1\Lambda \to K_1\Lambda'$. Thus K_1 is a covariant functor.

REMARK. Let P be any finitely generated projective over Λ, and let Aut(P) denote the group of Λ-linear automorphisms of P. It is interesting to note that there is a canonical homomorphism

$$\mathrm{Aut}(P) \to K_1\Lambda.$$

In the classical case, linear transformations of a vector space, this turns out to be just the determinant homomorphism.

The construction is as follows. Choose a finitely generated projective Q so that $P \oplus Q$ is free, and choose a basis $b_1, \ldots, b_r$ for $P \oplus Q$ over Λ. Each automorphism α of P gives rise to an automorphism $\alpha \oplus 1_Q$ of $P \oplus Q$. Using the chosen basis, this automorphism is represented by a matrix in the group $GL(r, \Lambda)$.

LEMMA 3.2. *The resulting embedding*

$$\mathrm{Aut}(P) \subset \mathrm{Aut}(P \oplus Q) \cong GL(r,\Lambda) \subset GL(\Lambda)$$

is well defined up to inner automorphism of $GL(\Lambda)$; *and hence gives rise to a well defined homomorphism* $\mathrm{Aut}(P) \to K_1\Lambda$.

Proof. Let $b_1', \ldots, b_s'$ be a different basis for $P \oplus Q$. (We must allow the possibility that $s \neq r$.) Then $b_i' = \Sigma\, c_{ij} b_j$ where the $s \times r$ matrix $C = (c_{ij})$ is invertible. Let A be the matrix of $\alpha \oplus 1_Q$ with respect to the original basis $\{b_j\}$. Conjugating A within $GL(r+s, \Lambda)$ by the square matrix $\begin{pmatrix} C & 0 \\ 0 & C^{-1} \end{pmatrix}$, we obtain the matrix $CAC^{-1} \in GL(s,\Lambda)$, which describes the automorphism $\alpha \oplus 1_Q$ with respect to the new basis. Thus a change of basis alters our embedding only by an inner automorphism of $GL(\Lambda)$. Now if we choose some other module Q' in place of Q, with say $P \oplus Q' \cong \Lambda^t$, then $Q \oplus \Lambda^t \cong Q' \oplus \Lambda^r$. Hence a different choice for Q can also alter our embedding only by an inner automorphism. This completes the proof. ■

Now suppose that the ring Λ is commutative. *Then a natural product operation*

$$K_0\Lambda \otimes K_1\Lambda \to K_1\Lambda$$

can be defined as follows. Let [P] be any generator of $K_0\Lambda$. The correspondence

$$\alpha \mapsto 1_P \otimes \alpha$$

defines a homomorphism from the automorphism group $\mathrm{Aut}(\Lambda^n) \cong \mathrm{GL}(n,\Lambda)$ to $\mathrm{Aut}(P \otimes \Lambda^n)$. (Here the symbol $\otimes$ stands for the tensor product over Λ.) Combining this construction with 3.2, we obtain a composite homomorphism

$$\mathrm{GL}(n,\Lambda) \to \mathrm{Aut}(P \otimes \Lambda^n) \to K_1\Lambda,$$

which will be called h(P). The identity

$$h(P \oplus P') = h(P) + h(P')$$

shows that the homomorphism h(P) depends only on the stable isomorphism class of P, and hence depends only on the element $[P] \in K_0\Lambda$. Now pass to the direct limit as $n \to \infty$, and abelianize. By definition, the resulting homomorphism from $K_1\Lambda$ to $K_1\Lambda$ carries each element k to the product $[P] \cdot k$. Thus we obtain a product operation $K_0\Lambda \otimes K_1\Lambda \to K_1\Lambda$, making $K_1\Lambda$ into a module over the ring $K_0\Lambda$.

Another distinctive feature of the commutative case is that the determinant operation is defined, and can be used to split $K_1\Lambda$ into a direct sum. Let $\Lambda^{\bullet}$ denote the multiplicative group consisting of all units of Λ. Then the composition

$$\Lambda^{\bullet} = \mathrm{GL}(1,\Lambda) \subset \mathrm{GL}(\Lambda) \xrightarrow{\det} \Lambda^{\bullet}$$

is evidently the identity map. Hence, if $\mathrm{SL}(\Lambda) \subset \mathrm{GL}(\Lambda)$ denotes the kernel of the determinant homomorphism, we obtain a direct sum decomposition

$$K_1\Lambda \cong \Lambda^{\bullet} \oplus (\mathrm{SL}(\Lambda)/E(\Lambda)).$$

The notation $SK_1\Lambda$ will sometimes be used for the second summand.

In many interesting cases the special linear group $SL(\Lambda)$ is generated by elementary matrices, and therefore $K_1\Lambda \cong \Lambda^{\bullet}$. This is the case, for example, if Λ is a field, or a local ring, or if Λ possesses a euclidean algorithm, or if Λ is the ring of integers in a finite extension of the rational numbers. (Compare §16.3.)

For further information and references see Milnor, *Whitehead torsion*, Bull. Amer. Math. Soc., 72 (1966), 358-426, or Kervaire, *Le groupe de Whitehead*, (notes by J. M. Arnaudies, mimeographed), Ecole Norm. Sup. Paris, 1966, or de Rham, Maumary, Kervaire, *Torsion et Type Simple d'Homotopie*, Springer Lecture Notes 48 (1967).

The "Mayer-Vietoris" Exact Sequence

Now consider a square of ring homomorphisms satisfying Hypotheses 1 and 2 of §2. We will construct an exact sequence

$$K_1\Lambda \to K_1\Lambda_1 \oplus K_1\Lambda_2 \to K_1\Lambda' \to K_0\Lambda \to K_0\Lambda_1 \oplus K_0\Lambda_2 \to K_0\Lambda'$$

of length six.

Define the homomorphisms

$$K_\alpha\Lambda \to K_\alpha\Lambda_1 \oplus K_\alpha\Lambda_2 \to K_\alpha\Lambda'$$

by

$$x \mapsto (i_{1*}\, x, i_{2*}\, x),$$

and

$$(y,z) \mapsto j_{1*}\, y - j_{2*}\, z$$

respectively. Define the homomorphism $\partial : K_1\Lambda' \to K_0\Lambda$ as follows. Represent the element x of $K_1\Lambda'$ by a matrix in $GL(n,\Lambda')$. This matrix determines an isomorphism h from the free Λ'-module $j_{1\#}\,\Lambda_1^n$ to the free Λ'-module $j_{2\#}\,\Lambda_2^n$. Hence, in the notation of §2, we can form the projective module $M = M(\Lambda_1^n, \Lambda_2^n, h)$ over Λ. Let

$$\partial(x) = [M] - [\Lambda^n] \in K_0\Lambda.$$

It is not difficult to verify that ∂ is a well defined homomorphism.

THEOREM 3.3. *The resulting sequence of length six is exact.*

The proof is not difficult. Details will be omitted.

There is an evident analogy between this sequence and the Mayer-Vietoris sequence of algebraic topology. (See for example Eilenberg and Steenrod, *Foundations of Algebraic Topology*, p. 39.) Hence we will refer to our sequence as a *Mayer-Vietoris sequence* also.

Example. Let Π be a cyclic group of prime order p with generator t, and let

$$\xi = e^{2\pi i/p}.$$

Then it is known that $Z[\xi]$ is the integral closure of Z in the cyclotomic field $Q[\xi]$, and hence is a Dedekind domain. (See §1.4. Compare Lang, *Algebraic Number Theory*, Addison-Wesley, 1970, p. 75.)

Define the square of homomorphisms

$$\begin{array}{ccc} Z\Pi & \xrightarrow{i_1} & Z[\xi] \\ \downarrow{\scriptstyle i_2} & & \downarrow{\scriptstyle j_1} \\ Z & \xrightarrow{j_2} & F_p \end{array}$$

by

$$i_1(t) = \xi, \quad j_1(\xi) = 1, \quad i_2(t) = 1.$$

Here $Z\Pi$ denotes the integral group ring of the cyclic group Π, and F_p denotes the field Z/pZ of integers modulo p. It is not difficult to verify that Hypotheses 1 and 2 are satisfied. Hence we obtain the exact sequence

$$\cdots \to K_1Z[\xi] \oplus K_1Z \to K_1F_p \to K_0Z\Pi \to K_0Z[\xi] \oplus K_0Z \to K_0F_p.$$

We will prove:

THEOREM OF RIM. *The homomorphism* $i_{1*} : K_0Z\Pi \to K_0Z[\xi]$ *is an isomorphism.*

Hence the computation of $K_0Z\Pi$ is reduced to a study of the ideal class group of the Dedekind domain $Z[\xi]$, where $\xi = e^{2\pi i/p}$. (See §1.11.)

REMARK 3.4. The precise structure of this ideal class group is known only for primes $p < 50$. It is shown in class field theory that the norm homomorphism

$$\text{ideal class group } (Z[\xi]) \to \text{ideal class group } (Z[\xi + \xi^{-1}])$$

is always surjective. (Compare C. Chevalley, C. R. Acad. Sci. Paris 192 (1931), 257-258, or K. Iwasawa, Abh. Math. Sem. Hamburg 20 (1956), 257-258. The domain $Z[\xi + \xi^{-1}]$ is precisely the ring of integers in the maximal real subfield of $Q(\xi)$.) The order of the kernel of this norm homomorphism is called the *relative class number* $h_1(p)$, and the order of the image is customarily denoted by $h_2(p)$.

Kummer, some 120 years ago, gave an explicit formula for the relative class number, and carried out numerical computations for $p < 100$, giving further details about the structure of the kernel in several special cases. (An excellent exposition of the classical theory may be found in Borevich and Shafarevich, *Number Theory*. For further references, see *Whitehead Torsion*, Bull. Amer. Math. Soc. 72, pp. 413, 419.) The second factor $h_2(p)$ of the class number is much more difficult to compute. However H. Bauer has recently shown that $h_2(p) = 1$ for $p < 50$. (Journal of Number Theory 1 (1969), 161-162.) Combining Bauer's statement with Kummer's explicit computations, we obtain the following table; where the cyclic group of order m is denoted briefly by (m).

p	ideal class group $(Z[\xi])$
≤ 19	(1)
23	(3)
29	$(2) \oplus (2) \oplus (2)$
31	(9)
37	(37)
41	$(11) \oplus (11)$
43	(211)
47	$(5) \oplus (139)$

The relative class number $h_1(p)$ grows very rapidly with p. Thus, for example

$$h_1(97) = 411,322,824,001.$$

Kummer, in 1851, claimed that $h_1(p)$ is asymptotically equal to the expression

$$G(p) = 2p(p/4\pi^2)^{(p-1)/4}$$

as $p \to \infty$. However this statement has never been proved, and may well be false. Ankeny and Chowla (Canadian J. Math. 3 (1951), 486-494) prove the weaker statement that

$$\log(h_1(p)/G(p)) = o(\log p)$$

as $p \to \infty$. (See also C. L. Siegel, *Zu zwei Bemerkungen Kummers,* Gesam. Abh. III, p. 440.) This implies that $h_1(p)$ is a strictly increasing function of p, for large values of p. The proof is based on the identity

$$\log(h_1(p)/G(p)) = \frac{p-1}{2} \sum \pm 1/nq^n,$$

to be summed over all prime powers q^n which are congruent to ± 1 modulo p, where each summand is positive or negative according as $q^n \equiv +1$ or $q^n \equiv -1$.

Much less is known about the second factor $h_2(p)$. Among primes less than 197, there is only one, namely 163, for which $h_2(p)$ is even. (See Kummer, Monatsber. K. Akad. Wiss. Berlin 1870, 855-880.) It has been conjectured that $h_2(p) = 1$ for $p < 97$, but that $h_2(97) \neq 1$. (Schrutka v. Rechtenstamm, Abh. D. Akad. Wiss. Berlin 2 (1964), p. 4.) For further information, see Ankeny, Chowla, and Hasse, *On the class number of the maximal real subfield of a cyclotomic field,* Journ. reine angew. Math. 217 (1965), 217-220, as well as C. Giffen, *Diffeotopically trivial periodic diffeomorphisms,* Invent. math. 11 (1970), 340-348.

Proof of Rim's theorem. Since $j_{2*} : K_0Z \to K_0F_p$ is clearly an isomorphism, it will suffice to verify that the homomorphism $j_{1*} : K_1Z[\xi] \to K_1F_p$ is surjective. But K_1F_p is just the group of units of F_p, or in other words the group of relatively prime residue classes modulo p.

For each integer k such that $1 \leq k \leq p-1$ consider the element

$$u = (\xi^k - 1)/(\xi - 1) = 1 + \xi + \dots + \xi^{k-1}$$

of the integral domain $Z[\xi]$. Setting $\xi^k = \eta$ and $\eta^\ell = \xi$ we can also consider the element

$$v = (\eta^\ell - 1)/(\eta - 1)$$

of $Z[\xi]$. Clearly $uv = 1$ so u is a unit of $Z[\xi]$. Since $j_1(u)$ is the residue class of k modulo p, this shows that j_{1*} is surjective, which completes the proof of Rim's theorem. (Compare Bass, *Algebraic K-Theory,* p. 604.) ∎

REMARK. A similar technique can be used to compute the group $K_1 Z\Pi$, using results to be established later in §6.4, §9.13, and §16.3. In fact our Mayer-Vietoris sequence extends to the left as follows

$$\dots \to 0 \to K_1 Z\Pi \to K_1 Z[\xi] \oplus K_1 Z \to K_1 F_p \overset{\partial}{\to} K_0 Z\Pi \to \dots .$$

Furthermore the group $K_1 Z[\xi]$ can be identified with the group of units of $Z[\xi]$. It follows easily that $K_1 Z\Pi$ can be identified with the group of units of $Z\Pi$; splitting as the direct sum of a finite group $\pm\Pi$ of order 2p and a free abelian group. Using the Dirichlet unit theorem, this free abelian summand is trivial for $p \leq 3$, but has rank $(p-3)/2 > 0$ if $p > 3$. (For further information and references, see Bass, *Algebraic K-Theory,* pp. 619-626.)

§4. The Exact Sequence Associated with an Ideal

Let $\mathfrak{a}$ be a two-sided ideal in the ring Λ. We will define groups $K_0\mathfrak{a}$ and $K_1\mathfrak{a}$ so as to prove the following.

LEMMA 4.1. *Each such ideal* $\mathfrak{a} \subset \Lambda$ *gives rise to an exact sequence*

$$K_1\mathfrak{a} \to K_1\Lambda \to K_1\Lambda/\mathfrak{a}$$
$$\to K_0\mathfrak{a} \to K_0\Lambda \to K_0\Lambda/\mathfrak{a}$$

of length six.

The definition of the group $K_i\mathfrak{a}$ follows.

By the *double* $D = D(\Lambda, \mathfrak{a})$ of the ring Λ along the ideal $\mathfrak{a}$ will be meant the subring of $\Lambda \times \Lambda$ consisting of all pairs (λ, λ') with $\lambda \equiv \lambda'$ mod $\mathfrak{a}$. Let $p_1 : D \to \Lambda$ and $p_2 : D \to \Lambda$ be the two projection maps,

$$p_1(\lambda, \lambda') = \lambda, \qquad p_2(\lambda, \lambda') = \lambda'.$$

DEFINITION (Bass, Stein). $K_i\mathfrak{a}$ is defined to be the kernel of the homomorphism $p_{1*} : K_iD \to K_i\Lambda$ induced by p_1. Here i can be either 0 or 1.

Now consider the commutative square of ring homomorphisms

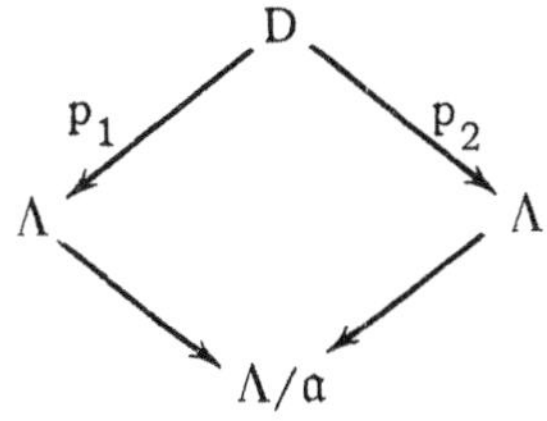

This clearly satisfies the hypotheses of §2, and hence gives rise to a Mayer-Vietoris exact sequence. (§3.3.) The required exact sequence

$K_1\mathfrak{a} \to K_1\Lambda \to K_1\Lambda/\mathfrak{a} \to K_0\mathfrak{a} \to K_0\Lambda \to K_0\Lambda/\mathfrak{a}$ can now be extracted from this Mayer-Vietoris exact sequence by inspection. (For example, as homomorphism $K_i\mathfrak{a} \to K_i\Lambda$ one uses the restriction to the subgroup (kernel p_{1*}) of the homomorphism $p_{2*} : K_iD \to K_i\Lambda$.) This completes the proof of 4.1. ∎

REMARK 1. If Λ is commutative, note that there is a product operation

$$K_0\Lambda \otimes K_i\mathfrak{a} \to K_i\mathfrak{a},$$

making $K_i\mathfrak{a}$ into a module over the ring $K_0\Lambda$. Similarly, a product operation

$$K_1\Lambda \otimes K_0\mathfrak{a} \to K_1\mathfrak{a}$$

is also defined. In fact, given elements

$$x \in K_i\Lambda \text{ and } y \in K_j\mathfrak{a} \subset K_jD,$$

with $i+j \leq 1$, we can use the diagonal embedding $\Delta : \Lambda \to D$ to form the product

$$(\Delta_* x) \cdot y \in K_{i+j}D;$$

and it is clear that this product actually belongs to the subgroup $K_{i+j}\mathfrak{a}$.

REMARK 2. The group K_iD is isomorphic to the direct sum $K_i\Lambda \oplus K_i\mathfrak{a}$. This is proved using the commutative triangle

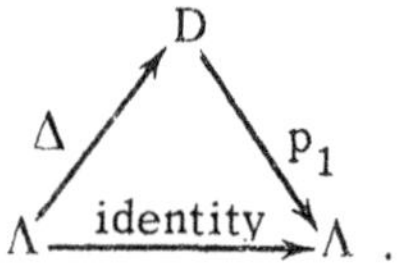

REMARK 3. If the ideal $\mathfrak{a}$ happens to be the entire ring Λ, then it seems at first that our notation is ambiguous. But in this case the natural map $K_i\mathfrak{a} \to K_i\Lambda$ is an isomorphism, so there is no problem. As another extreme example, if $\mathfrak{a}$ is the zero ideal then $K_i\mathfrak{a} = 0$.

REMARK 4. A more serious ambiquity occurs if $\mathfrak{a}$ can be considered as an ideal in more than one ring. Actually, one can show that the group $K_0\mathfrak{a}$ depends only on the structure of $\mathfrak{a}$ as a ring (without unit), so that $K_0\mathfrak{a}$ is independent of the ambient ring Λ; but the corresponding statement for $K_1\mathfrak{a}$ is false. (See Swan, *Excision in algebraic K-theory*, to appear; as well as §6.3.) Due caution should be exercised. Bass uses the notation $K_1(\Lambda, \mathfrak{a})$ for our $K_1\mathfrak{a}$ in order to emphasize the dependence on Λ. (*Algebraic K-Theory*, pp. 447, 369.)

Congruence Subgroups

Now let us study the group $K_1\mathfrak{a}$ from a quite different point of view. The kernel of the natural homomorphism

$$GL(n, \Lambda) \to GL(n, \Lambda/\mathfrak{a})$$

will be denoted by $GL(n, \mathfrak{a})$ and called the *congruence subgroup* determined by $\mathfrak{a}$. It consists of all invertible $n \times n$ matrices of the form $I + A$, where A is a matrix with entries in $\mathfrak{a}$.

Following Bass, consider the subgroup of $GL(n, \mathfrak{a})$ generated by all expressions of the form STS^{-1} where T is an elementary matrix belonging to $GL(n, \mathfrak{a})$ and S is an arbitrary element of the group $E(n, \Lambda)$ generated by $n \times n$ elementary matrices. We will use the notation $E(n, \mathfrak{a})$ for this subgroup. Thus $E(n, \mathfrak{a})$ is the smallest normal subgroup of $E(n, \Lambda)$ which contains every elementary matrix whose non-zero off-diagonal entry belongs to the ideal $\mathfrak{a}$. Passing to the direct limit as $n \to \infty$, we obtain corresponding groups $E(\mathfrak{a}) \subset GL(\mathfrak{a})$.

LEMMA 4.2. *The subgroup* $E(\mathfrak{a}) \subset GL(\mathfrak{a})$ *is normal, and the quotient* $GL(\mathfrak{a})/E(\mathfrak{a})$ *is canonically isomorphic to* $K_1\mathfrak{a}$.

Proof. Again let D denote the double of Λ along $\mathfrak{a}$. Clearly $GL(D)$ can be identified with the subgroup of $GL(\Lambda) \times GL(\Lambda)$ consisting of all pairs of invertible matrices which are congruent modulo $GL(\mathfrak{a})$. Notice that the subgroup

$$I \times GL(\mathfrak{a}) \subset GL(D),$$

where I denotes the identity matrix, is precisely the kernel of

$$p_{1*} : GL(D) \to GL(\Lambda).$$

Similarly notice that the subgroup $I \times E(\mathfrak{a})$ is precisely the kernel of

$$p_{1*} : E(D) \to E(\Lambda).$$

In fact for each generator STS^{-1} of $E(\mathfrak{a})$ the pair (I, STS^{-1}) can be written as $(S,S)\,(I,T)\,(S,S)^{-1}$, and therefore belongs to $E(D)$. Conversely if (I,X) is a product of elementary matrices in $GL(D)$, say

$$(I,X) = (S_1, S_1T_1) \ldots (S_k, S_kT_k),$$

then X is equal to the product

$$(S_1T_1S_1{}^{-1})\,(S_1S_2T_2S_2{}^{-1}S_1{}^{-1}) \ldots (S_1 \ldots S_kT_kS_k{}^{-1} \ldots S_1{}^{-1}),$$

and therefore belongs to $E(\mathfrak{a})$.

Now, by inspecting the commutative diagram

$$\begin{array}{ccccccccc}
1 & \to & E(\mathfrak{a}) & \to & GL(\mathfrak{a}) & \cdots\to & K_1\mathfrak{a} & \cdots\to & 1 \\
 & & \downarrow I\times & & \downarrow I\times & & \downarrow & & \\
1 & \to & E(D) & \to & GL(D) & \to & K_1D & \to & 1 \\
 & & \downarrow & & \downarrow & & \downarrow & & \\
1 & \to & E(\Lambda) & \to & GL(\Lambda) & \to & K_1\Lambda & \to & 1\ ,
\end{array}$$

we easily verify that the quotient group $GL(\mathfrak{a})/E(\mathfrak{a})$ is defined and isomorphic to $K_1\mathfrak{a}$. (Note that $E(D)$ maps onto $E(\Lambda)$.) This proves 4.2. ∎

The rest of §4 will be occupied with miscellaneous remarks which will not be used in subsequent sections.

Here is another characterization of $E(\mathfrak{a})$.

LEMMA 4.3 (Bass). *The group* $E(\mathfrak{a})$ *is equal to* $[E(\Lambda), E(\mathfrak{a})] = [GL(\Lambda), GL(\mathfrak{a})]$ *and hence is a normal subgroup of* $GL(\Lambda)$. *The center of the quotient group* $GL(\Lambda)/E(\mathfrak{a})$ *is precisely equal to* $GL(\mathfrak{a})/E(\mathfrak{a}) \cong K_1\mathfrak{a}$.

Here $[G_1, G_2]$ denotes the group generated by all commutators $g_1g_2g_1{}^{-1}g_2{}^{-1}$ with $g_1 \in G_1$, $g_2 \in G_2$.

Proof. The inclusion $[E(\Lambda), E(\mathfrak{a})] \subset E(\mathfrak{a})$ is valid since $E(\mathfrak{a})$ is a normal subgroup of $E(\Lambda)$; and equality holds since $E(\mathfrak{a})$ is generated by expressions of the form

$$STS^{-1} = [S,T]T,$$

where the elementary matrix T can be expressed as the commutator [S′, T′] of an elementary matrix and an elementary matrix in GL(𝔞). (Compare §5.)

For any matrix $X = I + A$ in GL(n,𝔞) note the identity

$$\begin{pmatrix} X & 0 \\ 0 & X^{-1} \end{pmatrix} = \begin{pmatrix} I & I \\ 0 & I \end{pmatrix}\begin{pmatrix} I & 0 \\ A & I \end{pmatrix}\begin{pmatrix} I & I \\ 0 & I \end{pmatrix}^{-1}\begin{pmatrix} I & X^{-1}A \\ 0 & I \end{pmatrix}\begin{pmatrix} I & 0 \\ -XA & I \end{pmatrix}$$

which shows that

$$\begin{pmatrix} X & 0 \\ 0 & X^{-1} \end{pmatrix} \in E(2n,\mathfrak{a}).$$

Since any generator [Y,X] of [GL(n,Λ), GL(n,𝔞)] can be expressed as the commutator of

$$\begin{pmatrix} Y & & \\ & I & \\ & & Y^{-1} \end{pmatrix} \text{ and } \begin{pmatrix} X & & \\ & X^{-1} & \\ & & I \end{pmatrix},$$

and hence belongs to [E(3n,Λ), E(3n,𝔞)], we obtain

$$[GL(n,\Lambda), GL(n,\mathfrak{a})] \subset [E(3n,\Lambda), E(3n,\mathfrak{a})].$$

Passing to the limit as $n \to \infty$, this shows that

$$[GL(\Lambda), GL(\mathfrak{a})] = [E(\Lambda), E(\mathfrak{a})].$$

Hence $E(\mathfrak{a}) = [GL(\Lambda), GL(\mathfrak{a})]$, which proves that E(𝔞) is normal in GL(Λ), and that the quotient GL(𝔞)/E(𝔞) is contained in the center of GL(Λ)/E(𝔞). But every element c in the center of GL(Λ)/E(𝔞) must map into an element c′ of the center of

$$GL(\Lambda)/GL(\mathfrak{a}) \subset GL(\Lambda/\mathfrak{a}).$$

This image c′ must commute with all elementary matrices in GL(Λ/𝔞), and hence must be equal to I. Therefore $c \in GL(\mathfrak{a})/E(\mathfrak{a})$, which completes the proof of 4.3. ■

Note that the problem of computing the groups $K_1(\mathfrak{a})$ is equivalent to the problem of determining all possible normal subgroups of GL(Λ). In fact, given a normal subgroup $N \subset GL(\Lambda)$, Bass has shown that there exists one and only one two-sided ideal 𝔞 so that $E(\mathfrak{a}) \subset N \subset GL(\mathfrak{a})$. (See Bass,

K-theory and stable algebra, §3.1.) Thus N determines a subgroup $N/E(\mathfrak{a}) \subset K_1(\mathfrak{a})$. Conversely, for any $\mathfrak{a}$ it is clear that any subgroup of $K_1(\mathfrak{a})$ determines a corresponding subgroup $N \subset GL(\mathfrak{a})$ which is normal in $GL(\Lambda)$.

Now suppose that Λ is a commutative ring. Then the determinant homomorphism $GL(\Lambda) \to \Lambda^\bullet$ is defined, and gives rise to a direct sum decomposition

$$K_1\mathfrak{a} = U(\mathfrak{a}) \oplus SK_1\mathfrak{a}.$$

Here $U(\mathfrak{a})$ denotes the group of units congruent to 1 modulo $\mathfrak{a}$, and

$$SK_1\mathfrak{a} = SL(\mathfrak{a})/E(\mathfrak{a}),$$

where $SL(\mathfrak{a})$ denotes the group of matrices with determinant 1 in the congruence subgroup $GL(\mathfrak{a})$.

If Λ is a Dedekind domain, then one can show that the group $SK_1(\mathfrak{a})$ is independent of $\mathfrak{a}$ at least to the following extent. If $\mathfrak{b} \subset \mathfrak{a}$ is a smaller non-zero ideal, then the natural homomorphism

$$SK_1(\mathfrak{b}) \to SK_1(\mathfrak{a})$$

is always surjective. (Compare §6.6.)

Examples. If Λ is the ring of integers Z, then $SK_1\mathfrak{a} = 0$ for every ideal $\mathfrak{a}$. (See J. Mennicke, *Finite factor groups of the unimodular group,* Annals of Math. 81 (1965), 31-37; or Bass, Lazard, and Serre, *Sous-groupes d'indice fini dans* SL(n,Z), Bull. Amer. Math. Soc. 70 (1964), 385-392.) The same is true if Λ is the ring of algebraic integers in any finite extension of the rationals which can be embedded in the field of real numbers. On the other hand, if Λ is the ring of integers in a totally imaginary number field, and if the ideal $\mathfrak{a}$ is sufficiently small, then $SK_1(\mathfrak{a})$ is a non-trivial finite cyclic group, isomorphic to the group of all roots of unity in Λ. For further information on these questions, the reader is referred to the papers listed at the end of the Preface.

§5. Steinberg Groups and the Functor K_2

This section will be concerned with relations between elementary matrices over a ring Λ.

Let $e_{ij}^{\lambda} \in GL(n,\Lambda)$ denote the elementary matrix with entry λ in the (i,j)-th place. Here i and j can be any distinct integers between 1 and n, and λ can be any ring element. The notation has been chosen so that

$$e_{ij}^{\lambda}\, e_{ij}^{\mu} = e_{ij}^{\lambda+\mu}.$$

(Note in particular that $(e_{ij}^{\lambda})^{-1} = e_{ij}^{-\lambda}$.)

The commutator $[a,b] = aba^{-1}b^{-1}$ of two elementary matrices can be computed as follows

$$[e_{ij}^{\lambda},\, e_{k\ell}^{\mu}] = \begin{cases} 1 & \text{if } j \neq k,\ i \neq \ell \\ e_{i\ell}^{\lambda\mu} & \text{if } j = k,\ i \neq \ell \\ e_{kj}^{-\mu\lambda} & \text{if } j \neq k,\ i = \ell. \end{cases}$$

(Note that this list does not include the case $j = k$, $i = \ell$. In fact there is no simple formula for $[e_{ij}^{\lambda},\, e_{ji}^{\mu}]$.)

Following R. Steinberg, *Générateurs, relations et revêtements de groupes algébriques*, Colloq. Théorie des groupes algébriques, Bruxelles 1962, 113-127, we introduce an abstract group defined by generators and relations which are designed to imitate the behavior of elementary matrices. Again let i,j range over all pairs of distinct integers between 1 and n and let λ and μ range over Λ.

DEFINITION. For $n \geq 3$ the *Steinberg group* $St(n,\Lambda)$ is the group defined by generators x_{ij}^{λ} subject to the relations

(1) $x_{ij}^{\lambda} x_{ij}^{\mu} = x_{ij}^{\lambda+\mu}$

(2) $[x_{ij}^{\lambda}, x_{j\ell}^{\mu}] = x_{i\ell}^{\lambda\mu}$ for $i \neq \ell$, and

(3) $[x_{ij}^{\lambda}, x_{k\ell}^{\mu}] = 1$ for $j \neq k$, $i \neq \ell$.

(In other words $St(n,\Lambda)$ is defined as a quotient $\mathfrak{F}/\mathfrak{R}$ where $\mathfrak{F}$ denotes the free group generated by the symbols x_{ij}^{λ} and $\mathfrak{R}$ denotes the smallest normal subgroup modulo which the above relations are valid.)

The restriction $n \geq 3$ is needed since these relations are completely inadequate when $n = 2$.

Define the *canonical homomorphism*

$$\phi : St(n,\Lambda) \to GL(n,\Lambda)$$

by the formula $\phi(x_{ij}^{\lambda}) = e_{ij}^{\lambda}$. This assignment does give rise to a homomorphism since each of the defining relations between generators of $St(n,\Lambda)$ maps into a valid identity between elementary matrices. The image $\phi(St(n,\Lambda)) \subset GL(n,\Lambda)$ is of course equal to the subgroup $E(n,\Lambda)$ generated by all elementary matrices.

Now pass to the direct limit as $n \to \infty$, thus obtaining corresponding groups and a corresponding homomorphism

$$\phi : St(\Lambda) \to GL(\Lambda).$$

Note that the image $\phi(St(\Lambda)) = E(\Lambda)$ is equal to the commutator subgroup of $GL(\Lambda)$. (Compare 3.1.)

DEFINITION. The kernel of the homomorphism $\phi : St(\Lambda) \to GL(\Lambda)$ will be called $K_2\Lambda$.

We will prove:

THEOREM 5.1. *The group* $K_2\Lambda$ = kernel (ϕ) *is precisely the center of the Steinberg group* $St(\Lambda)$.

Thus $K_2\Lambda$ is an abelian group which fits into the exact sequence

$$1 \to K_2\Lambda \to St(\Lambda) \to GL(\Lambda) \to K_1\Lambda \to 1.$$

Intuitively speaking we may think of $K_2\Lambda$ as the set of all nontrivial relations between elementary matrices,* the consequences of relations (1), (2), and (3) being the "trivial" relations. In fact any relation

$$e_{i_1j_1}^{\lambda_1}\, e_{i_2j_2}^{\lambda_2} \cdots e_{i_rj_r}^{\lambda_r} = I$$

between elementary matrices gives rise to an element $x_{i_1j_1}^{\lambda_1}\, x_{i_2j_2}^{\lambda_2} \cdots x_{i_rj_r}^{\lambda_r}$ of $K_2\Lambda$, and every element of $K_2\Lambda$ can be obtained in this way.

As an example the matrix

$$e_{12}^{1}\, e_{21}^{-1}\, e_{12}^{1} = \begin{pmatrix} 0 & 1 \\ -1 & 0 \end{pmatrix}$$

in $E(2,Z)$ represents a 90° rotation, and hence has period 4. The relation

$$(e_{12}^{1}\, e_{21}^{-1}\, e_{12}^{1})^4 = I$$

in $E(Z)$ gives rise to an element $(x_{12}^{1}\, x_{21}^{-1}\, x_{12}^{1})^4$ in K_2Z. We will see in §10 that the group K_2Z is cyclic of order 2, generated by this element $(x_{12}^{1}\, x_{21}^{-1}\, x_{12}^{1})^4$.

Note that K_2 is a covariant functor from rings to abelian groups. In fact every ring homomorphism $\Lambda \to \Lambda'$ clearly gives rise to a commutative diagram

$$\begin{array}{ccccccccc} 1 & \to & K_2\Lambda & \to & St(\Lambda) & \to & GL(\Lambda) & \to & K_1\Lambda & \to 1 \\ & & \downarrow & & \downarrow & & \downarrow & & \downarrow & \\ 1 & \to & K_2\Lambda' & \to & St(\Lambda') & \to & GL(\Lambda') & \to & K_1\Lambda' & \to 1 \end{array}$$

Proof of Theorem 5.1. First recall the well known fact that an $n \times n$ matrix (a_{ij}) commutes with every $n \times n$ elementary matrix $e_{k\ell}^{\lambda}$ if and only if (a_{ij}) is a diagonal matrix, with $a_{11} = a_{22} = \cdots = a_{nn}$ belonging to the center of Λ. For if (a_{ij}) commutes with $e_{k\ell}^{1}$ then direct computation shows that $a_{k\ell} = 0$ and $a_{kk} = a_{\ell\ell}$.

In particular note that no element of the subgroup

$$E(n-1, \Lambda) \subset E(n, \Lambda),$$

* Compare the thesis of S. Gersten, Cambridge, 1965.

other than I, belongs to the center of $E(n,\Lambda)$. Passing to the direct limit as $n \to \infty$, it follows that the limit group $E(\Lambda)$ has trivial center.

Now if c belongs to the center of $St(\Lambda)$ then $\phi(c)$ belongs to the center of $E(\Lambda)$, hence $\phi(c) = I$.

Conversely if $\phi(y) = I$ we must prove that y commutes with every generator x_{ij}^{λ} of the Steinberg group. Choose an integer n large enough so that y can be expressed as a word in the generators x_{ij}^{λ} with $i < n$ and $j < n$. Let P_n denote the subgroup of $St(\Lambda)$ generated by the elements $x_{1n}^{\mu}, x_{2n}^{\mu}, \ldots,$ and $x_{n-1,n}^{\mu}$, where μ ranges over Λ. This group is commutative by Relation (3).

LEMMA 5.2. *Each element of* P_n *can be written uniquely as a product*

$$x_{1n}^{\mu_1}\, x_{2n}^{\mu_2} \cdots x_{n-1,n}^{\mu_{n-1}} .$$

Hence the canonical homomorphism ϕ *maps* P_n *isomorphically into the group* $E(\Lambda)$.

The proof is immediate.

Next note that $x_{ij}^{\lambda}\, P_n\, x_{ij}^{-\lambda} \subset P_n$, providing that $i,j < n$. For each conjugate $x_{ij}^{\lambda}\, x_{kn}^{\mu}\, x_{ij}^{-\lambda}$ of a generator of P_n is equal either to x_{kn}^{μ} or to $x_{in}^{\lambda\mu}\, x_{kn}^{\mu}$ according as $j \neq k$ or $j = k$. But both of these expressions belong to P_n.

Now since y is a product of x_{ij}^{λ} with $i,j < n$ it follows that $y\, P_n\, y^{-1} \subset P_n$. But $\phi(y) = I$, so y actually commutes with each element p of P_n. For

$$\phi(y\, p\, y^{-1}) = \phi(y)\, \phi(p)\, \phi(y^{-1}) = \phi(p),$$

hence $y\, p\, y^{-1} = p$ by 5.2.

Therefore y commutes with every generator x_{kn}^{μ}, $k < n$. But a completely analogous argument shows that y also commutes with every $x_{n\ell}^{\mu}$, $\ell < n$. Hence y commutes with the commutator

$$[x_{kn}^{\mu}, x_{n\ell}^{1}] = x_{k\ell}^{\mu}$$

for all $k \neq \ell$; $k, \ell < n$. Since n can be arbitrarily large, this completes the proof of 5.1. ■

Universal Central Extensions

The Steinberg group $St(\Lambda)$ can be described from a more invariant point of view as the "universal central extension" of the group $E(\Lambda) = [GL(\Lambda), GL(\Lambda)]$. This description is due to Steinberg and Kervaire. Here are the relevant definitions.

By a *central extension* of a group G is meant a pair (X, ϕ) consisting of a group X and a homomorphism ϕ from X onto G with

$$\text{kernel}(\phi) \subset \text{center}(X).$$

(If the homomorphism ϕ is clear from the context we may, by abuse of language, refer to the group X as a central extension of G.)

A central extension (U, υ) of G is called *universal* if, for every central extension (X, ϕ) of G, there exists one and only one homomorphism from U to X over G. (That is, there exists one and only one $h : U \to X$ satisfying $\phi \circ h = \upsilon$.)

Clearly, if such a universal central extension exists, then it is unique up to isomorphism over G. In order to characterize universal central extensions we will need two further definitions.

A central extension (X, ϕ) of G *splits* if it admits a *section*, that is a homomorphism $s : G \to X$ over G. If (X, ϕ) splits, then clearly X is isomorphic to the cartesian product $G \times \text{kernel}(\phi)$.

A group G is *perfect* if G is equal to its commutator subgroup $[G, G]$.

THEOREM 5.3. *A central extension* (U, υ) *of* G *is universal if and only if* U *is perfect, and every central extension of* U *splits.*

Proof. First suppose that every central extension of U splits. Given a central extension (X, ϕ) of G, a homomorphism from U to X over G can be constructed as follows. Let $U \underset{G}{\times} X$ be the subgroup of $U \times X$ consisting of all (u, x) with $\upsilon(u) = \phi(x)$, and let $\pi(u, x) = u$.

Then $(U \times_G X, \pi)$ is a central extension of U, and hence possesses a section $s : U \to U \times_G X$. Setting $s(u) = (u, h(u))$ we obtain the required homomorphism h from U to X over G.

The proof that h is unique will be based on the following. Let (X, ϕ) and (Y, ψ) be central extensions of G.

LEMMA 5.4. *If* Y *is perfect, there exists at most one homomorphism from* Y *to* X *over* G.

Proof. Let f_1 and f_2 be homomorphisms from Y to X over G. Then for any y and z in Y we can write

$$f_1(y) = f_2(y)c, \quad f_1(z) = f_2(z)c',$$

where the elements c and c′ belong to the kernel of ϕ, and hence to the center of X. Therefore

$$f_1(yzy^{-1}z^{-1}) = f_2(yzy^{-1}z^{-1}).$$

Since Y is generated by commutators, this proves that $f_1 = f_2$. ■

Thus if U is perfect, and every central extension of U splits, we have proved that (U, υ) is a universal central extension of G. To prove the converse, we start with the converse of 5.4. Again let (Y, ψ) be a central extension of G.

LEMMA 5.5. *If* Y *is not perfect, then for suitably chosen* (X, ϕ) *there exists more than one homomorphism from* Y *to* X *over* G.

For, if Y is not perfect, there exists a non-zero homomorphism h from Y to some abelian group A. Let (X, ϕ) be the split extension $G \times A \to G$, with $\phi(g, a) = g$. Setting

$$f_1(y) = (\psi(y), 1), \quad f_2(y) = (\psi(y), h(y)),$$

we obtain two distinct homomorphisms from Y to $G \times A$ over G; thus proving 5.5. ■

We will also need the following.

LEMMA 5.6. *If* (X, ϕ) *is a central extension of a perfect group* P, *then the commutator subgroup* $X' = [X, X]$ *is perfect, and maps onto* P.

Thus $(X', \phi|X')$ is a perfect central extension of P.

Proof. Since P is generated by commutators, it is clear that ϕ maps X' onto P. Thus every element x of X can be written as a product $x'c$ with $x' \in X'$ and c central. Therefore every generator $[x_1, x_2]$ of X' is equal to $[x_1'c_1, x_2'c_2] = [x_1', x_2']$ for some x_1' and x_2' in X'. Thus $X' = [X', X']$, which proves 5.6. ■

We return to the proof of 5.3. Let (U, υ) be a universal central extension of G. It follows from 5.5 that U is perfect. We must show that every central extension of U splits.

Given a central extension (X, ϕ) of U, we will show that the composition

$$X \overset{\phi}{\to} U \overset{\upsilon}{\to} G$$

is a central extension of G. For if $\upsilon(\phi(x_0)) = 1$, then $\phi(x_0)$ belongs to the center of U, hence the correspondence

$$x \mapsto x_0 x x_0^{-1}$$

defines a homomorphism from X to X over U. Restricting to the commutator subgroup X', we see from 5.6 and 5.4 that the resulting homomorphism $X' \to X'$ over U is the identity. Thus x_0 commutes with elements of X'. But X is generated by X' together with the central subgroup kernel(ϕ), so it follows that x_0 commutes with all elements of X.

Thus $(X, \upsilon \circ \phi)$ is a central extension of G. Since (U, υ) is universal, there exists a homomorphism $s : U \to X$ over G. The composition $\phi \circ s$ is a homomorphism from U to U over G, hence equals the identity. Thus s is a section of (X, ϕ). This shows that every central extension of U splits, and completes the proof of 5.3. ■

THEOREM 5.7. *A group* G *admits a universal central extension if and only if* G *is perfect.*

For if (U, v) is a universal central extension, then U is perfect, hence its homomorphic image G must certainly be perfect.

Conversely, if G is perfect choose a homomorphism from a free group F onto G, and let $R \subset F$ denote the kernel. Then $[R, F]$ is a normal subgroup of F, and there is a natural surjection

$$\phi : F/[R, F] \to F/R \cong G.$$

Clearly the kernel of ϕ is central. So, by 5.6, the commutator subgroup

$$(F/[R, F])' \cong [F, F]/[R, F]$$

is a perfect central extension of G.

We claim that this central extension $[F, F]/[R, F] \to G$ is universal. Given any central extension (X, ψ) of G, since F is free there exists a homomorphism $h : F \to X$ over G. Furthermore $h([R, F]) = 1$, since (X, ψ) is central. Thus h induces a homomorphism from $F/[R, F]$ to X over G. Restricting to $[F, F]/[R, F]$, we obtain the required homomorphism from $[F, F]/[R, F]$ to X over G, which is unique by 5.4. ∎

COROLLARY 5.8. *The kernel of the universal central extension* $[F, F]/[R, F] \to G$ *is canonically isomorphic to the second homology group* $H_2(G; Z)$.

Proof. The kernel is clearly the central subgroup

$$(R \cap [F, F])/[R, F]$$

of $[F, F]/[R, F]$. But this expression is precisely the one which was used by Hopf in order to define the second homology group of G in the paper which pointed the way towards the modern field of homological algebra. (See H. Hopf, *Fundamentalgruppe und zweite Bettische Gruppe*, Comment. Math. Helv. 14 (1941-42), 257-309.) Nowadays we would note that the homology spectral sequence of the group extension $1 \to R \to F \to G \to 1$ gives rise to an exact sequence

$$H_2F \to H_2G \to H_0(G; H_1R) \to H_1F \to H_1G \to 0,$$

or in other words

$$0 \to H_2G \to R/[R, F] \to F/[F, F] \to 0 \to 0;$$

from which the conclusion follows. (I am indebted to Tate for this form of the argument.) Here and subsequently the notation H_2G always stands for $H_2(G; Z)$, where G operates trivially on Z. ■

REMARK. The universal central extension was first introduced, in the case of a finite group G, by J. Schur, *Über die Darstellung der endlichen Gruppen durch gebrochene lineare Substitutionem,* J. reine angew. Math. 127 (1904), 20-40. Hence the central subgroup

$$(R \cap [F, F])/[R, F] \cong H_2G$$

is often called the *Schur multiplier* (or Multiplikator) of G.

Here is a collection of familiar properties.

LEMMA 5.9. *Any homomorphism* $h : G_1 \to G_2$ *between perfect groups gives rise to a homomorphism* $h_* : H_2G_1 \to H_2G_2$ *between Schur multipliers. If* h *is an inner automorphism of* G, *then* h_* *is the identity map of* H_2G. *If* G *is the direct limit of a sequence* $G_1 \to G_2 \to G_3 \to \dots$ *of perfect groups, then* H_2G *is the direct limit of the sequence* $H_2G_1 \to H_2G_2 \to H_2G_2 \to \dots$.

Proofs are easily supplied.

Now consider the Steinberg group $St(\Lambda)$ associated with a ring Λ. It is clear from the defining relations that $St(\Lambda)$ is a perfect group; and it follows from 5.1 that $St(\Lambda)$ is a central extension of $E(\Lambda)$.

THEOREM 5.10. *The Steinberg group* $St(\Lambda)$ *is actually the universal central extension of* $E(\Lambda)$. *Hence the kernel* $K_2\Lambda$ *of the canonical map* $\phi : St(\Lambda) \to E(\Lambda)$ *can be identified with the Schur multiplier* $H_2E(\Lambda)$.

(See Steinberg, *Lectures on Chevalley groups*, p. 93, and Kervaire, *Multiplicateurs de Schur et* K*-theorie*, as well as Swan, *Algebraic* K*-theory*, p. 208.)

First an outline of the proof. Consider any central extension

$$(4) \qquad 1 \to C \to Y \overset{\phi}{\to} St(\Lambda) \to 1.$$

Then a section $s : St(\Lambda) \to Y$ is constructed as follows. For each generator x_{ij}^{λ} of $St(\Lambda)$ choose an index h distinct from i and j, choose

$$y \in \phi^{-1}(x_{ih}^{1}), \quad y' \in \phi^{-1}(x_{hj}^{\lambda}),$$

and form the commutator

$$s_{ij}^{\lambda} = [y, y'] \in \phi^{-1}(x_{ij}^{\lambda}).$$

This commutator clearly does not depend on the choice of y and y'. (Compare §8.) We will prove that it does not depend on the choice of h, and that these elements s_{ij}^{λ} satisfy all of the Steinberg relations. Thus the correspondence $x_{ij}^{\lambda} \mapsto s_{ij}^{\lambda}$ gives rise to the required homomorphism $s : St(\Lambda) \to Y$. Hence every central extension of $St(\Lambda)$ splits, and it follows from 5.3 that $(St(\Lambda), \phi)$ is the universal central extension of $E(\Lambda)$.

REMARK. The proof will apply equally well to central extensions of $St(n, \Lambda)$, providing that $n \geq 5$. Thus, if the kernel of the homomorphism $St(n, \Lambda) \to E(n, \Lambda)$ is central, and if $n \geq 5$, then $St(n, \Lambda)$ is the universal central extension $E(n, \Lambda)$. For example if Λ is a field or a skew-field then this is indeed the case. (Compare §9.12.) If Λ is a field, Steinberg has shown that $St(n, \Lambda)$ is the universal central extension of $E(n, \Lambda)$ even for $n = 3$ and $n = 4$, providing that one excludes just three exceptional cases: namely the groups $St(3, F_2)$, $St(3, F_4)$, and $St(4, F_2)$, which do have non-trivial Schur multipliers. (Here F_q denotes the field with q elements.)

To begin the proof of 5.10, consider any central extension

$$1 \to C \to Y \overset{\phi}{\to} St(n, \Lambda) \to 1$$

with $n \geq 5$. Given x and x' in $St(n, \Lambda)$, the symbol

$$[\phi^{-1}x, \phi^{-1}x'] \in Y$$

will denote the commutator $[y, y']$ where $y \in \phi^{-1}(x)$, $y' \in \phi^{-1}(x')$. Clearly this commutator does not depend on the choice of y and y'. We will be particularly interested in commutators of the form $[\phi^{-1}x_{ij}^{\lambda}, \phi^{-1}x_{k\ell}^{\mu}]$.

LEMMA 5.11. *If* $j \neq k$ *and* $\ell \neq i$ *then the commutator* $[\phi^{-1}x_{ij}^{\lambda}, \phi^{-1}x_{k\ell}^{\mu}]$ *is trivial.*

Proof. Since $n \geq 5$, we can choose an index h distinct from i, j, k, ℓ. Choosing

$$y \in \phi^{-1}(x_{ih}^{1}),\ y' \in \phi^{-1}(x_{hj}^{\lambda}),\ y'' \in \phi^{-1}(x_{k\ell}^{\mu}),$$

we have

$$[y, y'] \in \phi^{-1}(x_{ij}^{\lambda}).$$

But y commutes with y'' up to a central element (i.e., $[y, y''] \in C$) and y' commutes with y'' up to a central element, so it follows easily that $[y, y']$ actually commutes with y''. Thus

$$[\phi^{-1}x_{ij}^{\lambda}, \phi^{-1}x_{k\ell}^{\mu}] = [[y, y'], y''] = 1$$

as asserted. ∎

To continue the proof, we will need several identities between commutators which are valid for elements u, v, w of an arbitrary group G. Recall the definition

$$[u, v] = uvu^{-1}v^{-1} = [v, u]^{-1}.$$

The equality

$$(5) \qquad [u, v][u, w] = [u, vw][v, [w, u]]$$

is easily verified. Using this, we will prove the Jacobi identity

$$(6) \qquad [u, [v, w]][v, [w, u]][w, [u, v]] \equiv 1 \mod G'',$$

where $G'' = [[G, G], [G, G]]$ denotes the second commutator subgroup. In fact, writing (5) in the form $[v, [w, u]] = [vw, u][u, v][u, w]$, the left side of (6) is equal to

$$[uv, w][w, u][w, v][vw, u][u, v][u, w][wu, v][v, w][v, u]$$

$$\equiv [uv, w][wu, v][vw, u] = 1 \mod G''.$$

If the elements u and w happen to commute, note that the Jacobi identity can be formulated as an associative law

(7) $$[[u, v], w] \equiv [u, [v, w]] \mod G''.$$

We return to the proof of 5.10. Choose four distinct indices h, i, j, k and choose

$$u \in \phi^{-1}(x^1_{hi}),\ v \in \phi^{-1}(x^\lambda_{ij}),\ w \in \phi^{-1}(x^\mu_{jk}).$$

It follows from 5.11 that $[u, w] = 1$. Further, if G denotes the subgroup of Y generated by u, v and w, then the commutator subgroup G′ is generated by elements lying in $\phi^{-1}(x^\lambda_{hj})$, $\phi^{-1}(x^{\lambda\mu}_{ik})$, and $\phi^{-1}(x^{\lambda\mu}_{hk})$, so it follows that $G'' = 1$. Therefore

$$[[u, v], w] = [u, [v, w]],$$

or in other words

(8) $$[\phi^{-1}x^\lambda_{hj}, \phi^{-1}x^\mu_{jk}] = [\phi^{-1}x^1_{hi}, \phi^{-1}x^{\lambda\mu}_{ik}].$$

Taking $\lambda = 1$, this identity proves that the element

$$s^\mu_{hk} = [\phi^{-1}x^1_{hj}, \phi^{-1}x^\mu_{jk}]$$

does not depend on the choice of j.

Proof that these elements s^μ_{hk} *satisfy the three Steinberg relations.* Since $s^\mu_{hk} \in \phi^{-1}(x^\mu_{hk})$, the identity (8) can be rewritten as

$$[s^\lambda_{hj}, s^\mu_{jk}] = s^{\lambda\mu}_{hk}, \quad \text{for } h, j, k \text{ distinct.}$$

To prove the first Steinberg relation, we apply the identity

(5) $$[u, v][u, w] = [u, vw][v, [w, u]]$$

to elements

$$u \in \phi^{-1}(x^1_{hj}),\ v \in \phi^{-1}(x^\lambda_{jk}),\ \text{and}\ w \in \phi^{-1}(x^\mu_{jk}).$$

Then

$$[u, v] = s^\lambda_{hk},\ [u, w] = s^\mu_{hk},\ [u, vw] = s^{\lambda+\mu}_{hk},$$

and

$$[v, [w, u]] = [\phi^{-1}x^\lambda_{jk}, \phi^{-1}x^{-\mu}_{hk}] = 1,$$

so we obtain the required relation

$$s_{hk}^{\lambda} s_{hk}^{\mu} = s_{hk}^{\lambda+\mu}.$$

But the third Steinberg relation follows from 5.11, so all three relations are satisfied. This proves that *every central extension of* St(n, Λ) *splits, for* $n \geq 5$; and completes the proof of 5.10. ∎

A Product Operation

Now suppose that Λ is commutative. Using the expression $K_2\Lambda \cong H_2\, E(\Lambda)$, let us prove that $K_2\Lambda$ is a module over the ring $K_0\Lambda$.

Proceeding as on p. 27, each finitely generated projective P, with $P \oplus Q \cong \Lambda^r$, gives rise to a homomorphism

$$h_P\colon GL(n,\Lambda) \to GL(rn,\Lambda) \subset GL(\Lambda)$$

which is well defined up to inner automorphism of GL(Λ). Clearly h_P carries E(n,Λ), $n \geq 3$, into the commutator subgroup E(Λ); and the resulting homomorphism

$$E(n,\Lambda) \to E(\Lambda)$$

is well defined up to inner automorphism of E(Λ). So the induced

$$h_{P*}: H_2E(n,\Lambda) \to H_2\, E(\Lambda)$$

is well defined. Furthermore it can be verified that $h_{(P\oplus Q)*} = h_{P*} + h_{Q*}$. Now passing to the limit as $n \to \infty$, we obtain the required homomorphism $[P]\cdot$ from $K_2\Lambda$ to $K_2\Lambda$.

Just as in §4 one can also define the group $K_2\mathfrak{a}$ of an ideal, and construct product operations

$$K_0\Lambda \otimes K_2\mathfrak{a} \to K_2\mathfrak{a}$$

and

$$K_2\Lambda \otimes K_0\mathfrak{a} \to K_2\mathfrak{a}.$$

§6. Extending the Exact Sequences

This section will extend the exact sequences of §3 and §4 by appropriate K_2 terms. The only noteworthy change is that we need an extra hypothesis in order to extend the Mayer-Vietoris sequence of §3: namely the hypothesis that all of the ring homomorphisms involved are surjective. We start with the sequence of §4.

Let $\mathfrak{a}$ be a two-sided ideal in the ring Λ. As in §4, consider the projection $p_1 : D \to \Lambda$ to the first factor, where D consists of all (λ,λ') with $\lambda \equiv \lambda' \bmod \mathfrak{a}$.

DEFINITION. The kernel of the induced homomorphism $p_{1*}: \mathrm{St}(D) \to \mathrm{St}(\Lambda)$ will be called $\mathrm{St}(\mathfrak{a})$.

LEMMA 6.1. *Mapping* $\mathrm{St}(\mathfrak{a})$ *to* $\mathrm{St}(\Lambda)$ *by* p_{2*}, *we obtain an exact sequence*

$$\mathrm{St}(\mathfrak{a}) \to \mathrm{St}(\Lambda) \to \mathrm{St}(\Lambda/\mathfrak{a}) \to 1.$$

Proof. Note first that $\mathrm{St}(\mathfrak{a})$ is equal to the subgroup of $\mathrm{St}(D)$ generated by all expressions of the form $\Delta_*(s)x_{ij}{}^{(0,a)}\Delta_*(s^{-1})$ with $s \in \mathrm{St}(\Lambda)$, and $a \in \mathfrak{a}$ so that $(0,a)$ is an element of the ring D. Here $\Delta : \Lambda \to D$ denotes the diagonal map. The proof of this statement is similar to the proof of 4.2, and will be left to the reader.

Hence the image $p_{2*}\mathrm{St}(\mathfrak{a})$ is the subgroup of $\mathrm{St}(\Lambda)$ generated by all $sx_{ij}^{a}s^{-1}$, or in other words is equal to the smallest normal subgroup containing x_{ij}^{a} for all $a \in \mathfrak{a}$. But if we adjoin to the defining generators and relations for the group $\mathrm{St}(\Lambda)$ the new relations $x_{ij}^{a} = 1$, and hence $x_{ij}^{\lambda} = x_{ij}^{\lambda'}$ whenever $\lambda \equiv \lambda' \bmod \mathfrak{a}$, then we clearly obtain precisely the defining generators and relations for the group $\mathrm{St}(\Lambda/\mathfrak{a})$. This completes the proof of 6.1. ■

Let us identify the congruence subgroup $GL(\mathfrak{a})$ with the kernel of the homomorphism $p_{1*}: GL(D) \to GL(\Lambda)$. Mapping the exact sequence

$$1 \to K_2D \to St(D) \to GL(D) \to K_1D \to 1$$

by p_{1*} onto the corresponding sequence for Λ, we obtain as kernel the exact sequence

$$1 \to K_2\mathfrak{a} \to St(\mathfrak{a}) \to GL(\mathfrak{a}) \to K_1\mathfrak{a} \to 1.$$

Here $K_2\mathfrak{a}$ is defined just as on p. 33. By inspection of the commutative diagram

$$\begin{array}{ccccccccccc} & & & & & & 1 & & & & \\ & & & & & & \downarrow & & & & \\ 1 & \to & K_2\mathfrak{a} & \to & St\mathfrak{a} & \to & GL\mathfrak{a} & \to & K_1\mathfrak{a} & \to & 1 \\ & & \downarrow & & \downarrow & & \downarrow & & \downarrow & & \\ 1 & \to & K_2\Lambda & \to & St\Lambda & \to & GL\Lambda & \to & K_1\Lambda & \to & 1 \\ & & \downarrow & & \downarrow & & \downarrow & & \downarrow & & \\ 1 & \to & K_2\Lambda' & \to & St\Lambda' & \to & GL\Lambda' & \to & K_1\Lambda' & \to & 1\,, \\ & & & & \downarrow & & & & & & \\ & & & & 1 & & & & & & \end{array}$$

where $\Lambda' = \Lambda/\mathfrak{a}$, we now prove the following.

THEOREM 6.2. *There is an exact sequence*

$$K_2\mathfrak{a} \to K_2\Lambda \to K_2\Lambda' \xrightarrow{\partial} K_1\mathfrak{a} \to K_1\Lambda \to \dots$$

extending the exact sequence of §4.

Here the homomorphism ∂ is obtained by proceeding from $K_2\Lambda'$ across to $St\Lambda'$, up to $St\Lambda$, across to $GL\Lambda$, up to $GL\mathfrak{a}$ (uniquely), and across to $K_1\mathfrak{a}$. This yields a well defined homomorphism, since any two elements of $St\Lambda$ with the same image in $St\Lambda'$ differ only by an element from $St\mathfrak{a}$ which maps into 1 in $K_1\mathfrak{a}$. Further details will be left to the reader. ∎

Next consider a ring homomorphism $f : \Lambda \to \Gamma$. Let $\mathfrak{a} \subset \Lambda$ and $\mathfrak{b} \subset \Gamma$ be two-sided ideals with $f(\mathfrak{a}) \subset \mathfrak{b}$. Then there is a corresponding commutative diagram

$$\begin{array}{ccccccccccc} 1 & \to & K_2\mathfrak{a} & \to & St\mathfrak{a} & \to & GL\mathfrak{a} & \to & K_1\mathfrak{a} & \to & 1 \\ & & \downarrow & & \downarrow & & \downarrow & & \downarrow & & \\ 1 & \to & K_2\mathfrak{b} & \to & St\mathfrak{b} & \to & GL\mathfrak{b} & \to & K_1\mathfrak{b} & \to & 1\,. \end{array}$$

EXCISION LEMMA 6.3. *If* $f : \Lambda \to \Gamma$ *is surjective, and maps the ideal* $\mathfrak{a}$ *one-to-one onto* $\mathfrak{b}$, *then the induced homomorphism* $f_*: K_2\mathfrak{a} \to K_2\mathfrak{b}$ *is surjective and* $f_*: K_1\mathfrak{a} \to K_1\mathfrak{b}$ *is an isomorphism.*

Proof. Since $\mathfrak{a}$ maps one-to-one onto $\mathfrak{b}$ it is clear that $GL(\mathfrak{a})$ maps isomorphically onto $GL(\mathfrak{b})$. Furthermore, recalling from 6.1 that $St(\mathfrak{b})$ is generated by expressions $\Delta_*(s)x_{ij}^{(0,b)}\Delta_*(s^{-1})$ with $s \in St(\Gamma)$, since $St(\Lambda)$ maps onto $St(\Gamma)$ and $\mathfrak{a}$ maps onto $\mathfrak{b}$ it is clear that $St(\mathfrak{a})$ maps onto $St(\mathfrak{b})$. The lemma now follows easily. (Compare Bass, *Algebraic K-Theory*, p. 382.) ■

Still assuming that f is surjective, and maps $\mathfrak{a}$ isomorphically onto $\mathfrak{b}$, we have the following commutative diagram:

$$\begin{array}{ccccccccccc}
K_2\mathfrak{a} & \to & K_2\Lambda & \to & K_2\Lambda' & \to & K_1\mathfrak{a} & \to & K_1\Lambda & \to & K_1\Lambda' \\
\downarrow{\scriptstyle \text{onto}} & & \downarrow & & \downarrow & & \downarrow{\scriptstyle \cong} & & \downarrow & & \downarrow \\
K_2\mathfrak{b} & \to & K_2\Gamma & \to & K_2\Gamma' & \to & K_1\mathfrak{b} & \to & K_1\Gamma & \to & K_1\Gamma' .
\end{array}$$

Here $\Lambda' = \Lambda/\mathfrak{a}$ and $\Gamma' = \Gamma/\mathfrak{b}$. By inspection of this diagram we easily obtain the exact sequence

$$K_2\Lambda \to K_2\Gamma \oplus K_2\Lambda' \to K_2\Gamma' \to K_1\Lambda \to K_1\Gamma \oplus K_1\Lambda' \to K_1\Gamma' .$$

Thus we have proved the following.

THEOREM 6.4. *If a commutative square of surjective ring homomorphisms*

$$\begin{array}{ccc}
\Lambda & \longrightarrow & \Lambda' \\
\downarrow & & \downarrow \\
\Gamma & \longrightarrow & \Gamma'
\end{array}$$

satisfies the hypotheses of §2, then the Mayer-Vietoris exact sequence of §3 can be extended by the terms

$$K_2\Lambda \to K_2\Gamma \oplus K_2\Lambda' \to K_2\Gamma' \to K_1\Lambda \to \dots .$$

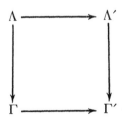

In fact the two hypotheses of §2 were that Λ should be equal to the product of Γ and Λ' over Γ', and that at least one of the two homomorphisms to Γ' should be surjective. But if $\Lambda' = \Lambda/\mathfrak{a}$ and $\Gamma' = \Gamma/\mathfrak{b}$ then the first hypothesis is clearly equivalent to the requirement that $\mathfrak{a}$ should map one-to-one onto $\mathfrak{b}$. ■

REMARK 6.5. Swan has recently shown that it is not possible to define any functor K_3 so that the exact sequences of 6.2 and 6.4 are extended by appropriate K_3-terms. (See Swan, *Excision in algebraic K-theory*, to appear.) This result suggests that our definition of $K_2\mathfrak{a}$ may not be too useful.

REMARK 6.6 To conclude this section, let us consider the rather different situation associated with a pair of two-sided ideals $\mathfrak{a} \subset \mathfrak{b}$ in the same ring Λ. The image of the ideal $\mathfrak{b}$ in the quotient ring $\Lambda/\mathfrak{a}$ will be denoted by $\mathfrak{b}/\mathfrak{a}$. Then it is not difficult to construct homomorphisms

$$(*) \qquad K_2\mathfrak{a} \to K_2\mathfrak{b} \to K_2\mathfrak{b}/\mathfrak{a} \to K_1\mathfrak{a} \to \dots ,$$

the composition of any two successive homomorphisms being zero. Consider the following commutative diagram of interlocked sequences

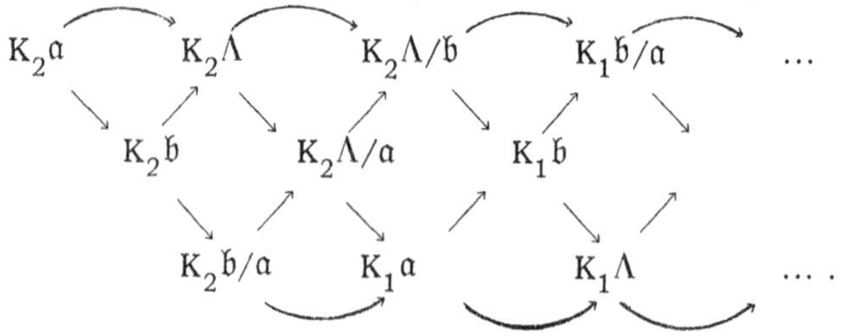

By inspection of this diagram one can verify that at least the following portion of the sequence $(*)$ is exact:

$$K_2\mathfrak{b}/\mathfrak{a} \to K_1\mathfrak{a} \to K_1\mathfrak{b} \to K_1\mathfrak{b}/\mathfrak{a} \to K_0\mathfrak{a} \to K_0\mathfrak{b} \to K_0\mathfrak{b}/\mathfrak{a}.$$

As a typical application, due to Bass, if Λ is a Dedekind domain and if $\mathfrak{a} \neq 0$, then it is not difficult to show that $SK_1\mathfrak{b}/\mathfrak{a} = 0$. So it follows that the homomorphism $SK_1\mathfrak{a} \to SK_1\mathfrak{b}$ must be surjective.

I do not know whether or not the initial portion of the sequence $(*)$ is exact.

§7. The Case of a Commutative Banach Algebra

Let Λ be a commutative Banach algebra over the real or complex numbers. For example Λ could be the ring R^X of continuous real valued functions on a compact space X. This section will use the topological structure of $GL(n,\Lambda)$ to compute $K_1\Lambda$ and to estimate $K_2\Lambda$.

LEMMA 7.1. *The group* $E(n,\Lambda)$ *is an open, path-connected subgroup of the special linear group* $SL(n,\Lambda)$.

Hence $E(n,\Lambda)$ is the component of the identity in $SL(n,\Lambda)$, and the quotient $SL(n,\Lambda)/E(n,\Lambda)$ can be identified with the group $\pi_0 SL(n,\Lambda)$ of path components.

Now recall that $K_1\Lambda$ splits as the direct sum of the group $\Lambda^\bullet$ of units and the group

$$\begin{aligned} SK_1(\Lambda) &= SL(\Lambda)/E(\Lambda) \\ &= \varinjlim SL(n,\Lambda)/E(n,\Lambda) = \varinjlim \pi_0\, SL(n,\Lambda). \end{aligned}$$

If we give $SL(\Lambda)$ the direct limit (= fine) topology, then clearly the group $\pi_0 SL(\Lambda)$ of path components can be identified with $\varinjlim \pi_0 SL(n,\Lambda)$. Thus we obtain:

COROLLARY 7.2. *The group* $K_1\Lambda$ *splits as the direct sum of the group* $\Lambda^\bullet$ *of units of* Λ *and the group* $SK_1(\Lambda) \cong \pi_0\, SL(\Lambda)$ *of path components of* $SL(\Lambda)$.

In the special case $\Lambda = R^X$, this description can be further simplified as follows. Since the multiplicative group $R^\bullet$ is topologically isomorphic to the additive group $F_2 \oplus R$, we have

$$(R^X)^\bullet \cong (R^\bullet)^X \cong (F_2)^X \oplus R^X.$$

Furthermore the special linear group $SL(n,R)$ contains the rotation group $SO(n)$ as deformation retract,* and hence the space

$$SL(n,R^X) \cong SL(n,R)^X$$

contains the function space $SO(n)^X$ as deformation retract. Passing to the limit as $n \to \infty$ we obtain:

COROLLARY 7.3. *The group* K_1R^X *splits as the direct sum of the group* $(R^\bullet)^X$ *of units and the group* $\pi_0(SO^X)$ *of homotopy classes of mappings from* X *to the infinite rotation group* SO.

Similarly, for the ring C^X of continuous complex valued functions, we obtain

$$K_1C^X \cong (C^\bullet)^X \oplus \pi_0(SU^X),$$

where SU denotes the infinite special unitary group.

Now let us prove Lemma 7.1. Since each elementary matrix e_{ij}^λ can be joined to I by the path

$$t \mapsto e_{ij}^{t\lambda}, \quad 0 \leq t \leq 1,$$

it is clear that $E(n,\Lambda)$ is path connected. To prove that $E(n,\Lambda)$ is open (and hence closed), we will prove a sharper statement. Let $I + A$ be an $n \times n$ matrix with determinant 1.

LEMMA 7.4. *If each entry* a_{ij} *of* A *satisfies* $\|a_{ij}\| < 1/(n-1)$ *then* $I + A$ *can be expressed canonically as a product of* $n^2 + 5n - 6$ *elementary matrices, each of which depends continuously on* A.

Proof. The entry $1 + a_{11}$ has an inverse u satisfying

$$\|u\| < \left(1 - \frac{1}{n-1}\right)^{-1} = \frac{n-1}{n-2}.$$

---

* This is proved, for example, using a form of the Gram-Schmidt orthonormalization process.

Subtracting ua_{k1} times the first row from the k-th for $k = 2, 3, \ldots, n$, we obtain a matrix $I + (a'_{kj})$ satisfying $a'_{21} = a'_{31} = \ldots = a'_{n1} = 0$ and

$$\begin{aligned} \|a'_{kj}\| &\leq \|a_{kj}\| + \|ua_{k1}\, a_{1j}\| \\ &< \frac{1}{n-1} + \frac{n-1}{n-2}\,\frac{1}{n-1}\,\frac{1}{n-1} = \frac{1}{n-2}. \end{aligned}$$

Continuing inductively, after $n(n-1)$ elementary row operations we have reduced the matrix to diagonal form. Let say

$$u^{-1},\ uv^{-1},\ vw^{-1},\ \ldots$$

be the diagonal entries. Multiplying by

$$e_{12}^{u}\, e_{21}^{-u^{-1}}\, e_{12}^{u}\, e_{12}^{-1}\, e_{21}^{1}\, e_{12}^{-1} = \begin{pmatrix} u & 0 \\ 0 & u^{-1} \end{pmatrix}$$

we can replace u by 1. Continuing inductively, after $6(n-1)$ row operations the diagonal matrix will be reduced to the identity. This proves Lemmas 7.4 and 7.1. ■

Note that the connected topological group $E(n,\Lambda)$ is locally contractible (this is easily proved using, for example, the exponential map) and hence has a universal covering group, which we will call $\tilde{E}$. This group $\tilde{E}$ is a central extension

$$1 \to \pi_1\, E(n,\Lambda) \to \tilde{E} \to E(n,\Lambda) \to 1.$$

For each elementary matrix e_{ij}^{λ} in $E(n,\Lambda)$ we will choose a representative $\tilde{e}_{ij}^{\lambda}$ in $\tilde{E}$ which depends continuously on λ and tends to the identity as $\lambda \to 0$. In fact define $\tilde{e}_{ij}^{\lambda}$ as the endpoint of the path which starts at 1 and covers the path $t \mapsto e_{ij}^{t\lambda}$.

Then these elements $\tilde{e}_{ij}^{\lambda}$ satisfy the Steinberg relations $\tilde{e}_{ij}^{\lambda}\, \tilde{e}_{ij}^{\mu} = \tilde{e}_{ij}^{\lambda+\mu}$; and $[\tilde{e}_{ij}^{\lambda}, \tilde{e}_{k\ell}^{\mu}] = 1$ or $\tilde{e}_{i\ell}^{\lambda\mu}$ according as $j \neq k$ or $j = k$, for $i \neq \ell$. This is proved by lifting the paths $e_{ij}^{t\lambda}\, e_{ij}^{t\mu} = e_{ij}^{t\lambda+t\mu}$ and $[e_{ij}^{\lambda}, e_{k\ell}^{t\mu}]$ respectively. Therefore the correspondence

$$\phi : x_{ij}^{\lambda} \mapsto \tilde{e}_{ij}^{\lambda}$$

defines a homomorphism from the Steinberg group $St(n,\Lambda)$ to $\tilde{E}$.

LEMMA 7.5. *This homomorphism* $\tilde{\phi} : \mathrm{St}(n,\Lambda) \to \tilde{E}$ *is surjective, and maps the kernel of* $\phi : \mathrm{St}(n,\Lambda) \to \mathrm{GL}(n,\Lambda)$ *onto the fundamental group*

$$\pi_1\ E(n,\Lambda) = \pi_1\ SL(n,\Lambda).$$

Proof. The explicit construction 7.4 can be used to define a continuous lifting of a neighborhood of I in $E(n,\Lambda)$ to an open subset of $\tilde{E}$, simply by replacing each e_{ij}^{λ} by $\tilde{e}_{ij}^{\lambda}$. Since any open subset of the connected group $\tilde{E}$ generates the group, this proves that $\tilde{\phi}$ is surjective. The rest is clear. ∎

Now taking the direct limit as $n \to \infty$ we obtain:

COROLLARY 7.6. *The limit homomorphism* $\tilde{\phi}$ *maps* $K_2\Lambda$ *onto* $\pi_1 SL(\Lambda)$.

As an example, if Λ is the Banach algebra of real numbers R, then the fundamental group

$$\pi_1\ SL(R) \cong \pi_1 SO$$

is cyclic of order 2. It follows that K_2R has at least two distinct elements.

Actually we will see in later sections that K_2R is the direct sum of a cyclic group of order 2, coming from K_2Z, and the group (kernel $\tilde{\phi}$) which is uncountable and infinitely divisible. (Compare §§8.4, 10.2, 11.10.) In particular, the surjection $\tilde{\phi} : K_2\Lambda \to \pi_1 SL(\Lambda)$ is certainly not an isomorphism when $\Lambda = R$.

Now suppose again that Λ is the ring R^X of continuous real valued functions on a compact space X. Then

$$\begin{aligned}\pi_1\ SL(n,R^X) &= \pi_1(SL(n,R)^X)\\ &= \pi_1(SO(n)^X)\end{aligned}$$

can be identified with the group of path components of the loop space $\Omega(SO(n)^X) = (\Omega\ SO(n))^X$. *Hence* $\pi_1\ SL(n,R^X)$ *is isomorphic to the group of homotopy classes of mappings from* X *to* $\Omega\ SO(n)$.

Taking the limit as $n \to \infty$, this proves:

COROLLARY 7.7. *The group* $K_2(R^X)$ *maps homomorphically onto the group of all homotopy classes of mappings from* X *to* Ω SO.

Example. If X is a 2-sphere, then this group of homotopy class is the sum of a cyclic group of order 2 (since Ω SO has two components) and an infinite cyclic group (since $\pi_2 \Omega \, SO \cong \pi_3 \, SO \cong Z$). *Therefore, in this case,* K_2R^X *maps onto an infinite cyclic group.*

Topological K-Theory

The functor K_0R^X has an analogous topological interpretation. If ξ is a real vector bundle over the compact space X, then the set $\Gamma(\xi)$ consisting of all continuous sections of ξ is a module over the ring R^X. Choosing η so that the Whitney sum $\xi \oplus \eta$ is a trivial vector bundle, it follows that $\Gamma(\xi) \oplus \Gamma(\eta)$ is free over R^X. Hence $\Gamma(\xi)$ is finitely generated and projective. Conversely, every finitely generated projective over R^X is isomorphic to $\Gamma(\xi)$ for some essentially unique vector bundle ξ. See R. Swan, *Vector bundles and projective modules,* Trans. Amer. Math. Soc. 105 (1962), 264-277.

Now consider the graded ring $KO^*(X)$ of topological K-theory. (See Atiyah, K-*Theory,* Benjamin 1967.) Evidently K_0R^X is canonically isomorphic to the ring $KO^0(X) \cong \pi_0((Z \times B_O)^X)$ of virtual real vector bundles over X. The conclusions of this section can be summarized briefly by the following exact sequence

$$\begin{array}{rlll} & K_2R^X \to & KO^{-2}X \overset{0}{\to} R^X \\ \overset{\exp}{\to} & K_1R^X \to & KO^{-1}X \to 0 \\ \to & K_0R^X \to & KO^0X \to 0 \, ; \end{array}$$

where the third homomorphism carries each $f : X \to R$ to the composition $\exp \circ f \in (R^\bullet)^X \subset K_1R^X$.

Similarly, for the ring C^X of continuous complex valued functions on X, there is an analogous exact sequence

$$\begin{aligned} & K_2 C^X \to K^{-2}X \overset{h}{\to} C^X \\ \overset{\exp}{\to}\ & K_1 C^X \to K^{-1}X \to 0 \\ \to\ & K_0 C^X \to K^0 X \quad \to 0. \end{aligned}$$

Here the homomorphism h can be defined as the composition of the natural surjection

$$K^{-2}X \cong K^0 X \to Z^X,$$

and the embedding

$$Z^X \cong (2\pi i Z)^X \subset C^X.$$

§8. The Product $K_1\Lambda \otimes K_1\Lambda \to K_2\Lambda$

Here is one procedure for constructing elements of $K_2\Lambda$. Given two matrices

$$A,B \in E(\Lambda)$$

which happen to commute, choose representatives $a,b \in St(\Lambda)$,

$$\phi(a) = A, \quad \phi(b) = B.$$

The commutator $[a,b] = aba^{-1}b^{-1}$ is then an element of $K_2\Lambda$. (Compare §5.10.) We will use the notation

$$A \star B = aba^{-1}b^{-1} \in K_2\Lambda$$

for this commutator.

Note that $A \star B$ does not depend on the choice of representatives a and b. If for example $\phi(\bar{a})$ is also equal to A, then $\bar{a} = ac$ for some element c in the center of $St(\Lambda)$, and hence

$$\bar{a}b\bar{a}^{-1}b^{-1} = a(cbc^{-1})a^{-1}b^{-1} = aba^{-1}b^{-1}.$$

LEMMA 8.1. *This construction is skew-symmetric*

$$B \star A = (A \star B)^{-1},$$

bimultiplicative

$$(A_1A_2) \star B = (A_1 \star B)(A_2 \star B),$$

and invariant under inner automorphism of $E(\Lambda)$

$$(PAP^{-1}) \star (PBP^{-1}) = A \star B.$$

Proof. These statements follow immediately from the commutator identities

$$[b,a] = [a,b]^{-1},$$

$$[a_1a_2,b] = [a_1,[a_2,b]]\,[a_2,b]\,[a_1,b]$$

and

$$[pap^{-1}, pbp^{-1}] = p[a,b]p^{-1}$$

respectively. ■

Now suppose that the ring Λ is commutative. Let u and v be any two units of Λ. Then the diagonal matrices

$$D_u = \begin{pmatrix} u & 0 & 0 \\ 0 & u^{-1} & 0 \\ 0 & 0 & 1 \end{pmatrix}, \quad D'_v = \begin{pmatrix} v & 0 & 0 \\ 0 & 1 & 0 \\ 0 & 0 & v^{-1} \end{pmatrix}$$

commute, and belong to $E(3,\Lambda)$.

DEFINITION. The commutator $D_u \star D'_v \in K_2\Lambda$ will be denoted briefly by $\{u,v\}$.

LEMMA 8.2. *The symbol* $\{u,v\}$ *is also skew-symmetric*

$$\{u,v\} = \{v,u\}^{-1},$$

and bimultiplicative

$$\{u_1u_2,v\} = \{u_1,v\}\{u_2,v\}.$$

Proof. Let

$$P = \begin{pmatrix} -1 & 0 & 0 \\ 0 & 0 & 1 \\ 0 & 1 & 0 \end{pmatrix}$$

in $E(3,\Lambda)$ so that

$$PD_uP^{-1} = D'_u, \qquad PD'_vP^{-1} = D_v.$$

Then

$$\begin{aligned} \{u,v\}^{-1} &= (D_u \star D'_v)^{-1} = D'_v \star D_u \\ &= (PD'_vP^{-1}) \star (PD_uP^{-1}) = D_v \star D'_u = \{v,u\}. \end{aligned}$$

The rest of the proof is clear. ■

Now we can completely describe $D \star D'$ for any pair of diagonal matrices in $SL(n,\Lambda)$. Let $\mathrm{diag}(u_1,\ldots,u_n)$ denote the diagonal matrix with diagonal entries $u_1,\ldots,u_n$.

LEMMA 8.3. *If* $u_1 u_2 \cdots u_n = v_1 v_2 \cdots v_n = 1$ *then*

$$\mathrm{diag}(u_1, \ldots, u_n) \star \mathrm{diag}(v_1, \ldots, v_n)$$

is equal to the product

$$\{u_1, v_1\}\{u_2, v_2\} \cdots \{u_n, v_n\}.$$

The proof will be left as an exercise.

Here are some examples to show that the symbol $\{u,v\}$ is not identically equal to 1. Recall from §7 that there is a natural homomorphism $\tilde{\phi}$ from K_2R to the fundamental group π_1 SL(R), which we will identify with the group consisting of $+1$ and -1.

LEMMA 8.4. *If* Λ *is the field* R *of real numbers, then* $\tilde{\phi}\{-1, -1\} = -1$.

It follows easily that

$$\tilde{\phi}\{u,v\} = -1 \text{ if } u < 0 \text{ and } v < 0,$$
$$\tilde{\phi}\{u,v\} = +1 \text{ if } u > 0 \text{ or } v > 0.$$

REMARKS. It follows that $\{-1, -1\} \neq 1$ whenever the ring Λ can be mapped homomorphically into the reals. In fact if Λ can be mapped to R in several different ways, then it usually follows that several different symbols $\{u,u\} \in K_2\Lambda$ are linearly independent modulo 2. For example if Λ is the ring $Z[\sqrt{2}]$ then $\{1+\sqrt{2}, 1+\sqrt{2}\}$ and $\{-1, -1\}$ are independent. Or if Λ is the group ring $Z\Pi$ where Π is cyclic of order 2 generated by t, then $\{t,t\}$ and $\{-1, -1\}$ are independent.

Proof of 8.4. The group SL(n,R) contains the rotation group SO(n) as deformation retract. Recall that the universal covering group of SO(3) can be identified with the group S^3 of unit quaternions. In fact if R_0^3 denotes the real vector space spanned by the unit quaternions i, j, k, then each $\xi \in S^3$ determines a rotation

$$\eta \mapsto \xi\eta\xi^{-1}$$

of R_0^3, and only the elements $\pm 1 \in S^3$ determine the identity rotation. Note in particular that the quaternion units k and j determine the rotations with matrices

$$D_{-1} = \begin{pmatrix} -1 & 0 & 0 \\ 0 & -1 & 0 \\ 0 & 0 & 1 \end{pmatrix} \text{ and } D'_{-1} = \begin{pmatrix} -1 & 0 & 0 \\ 0 & 1 & 0 \\ 0 & 0 & -1 \end{pmatrix}$$

respectively. Hence

$$\tilde{\phi}\{-1, -1\} = \tilde{\phi}(D_{-1} \star D'_{-1}) = kjk^{-1}j^{-1} = -1.$$

This completes the proof. ■

Here is a more delicate example using the techniques of §7. Let Λ be any commutative ring which can be mapped to the complex numbers, and let u and v be indeterminates. Then we can form the ring $\Lambda[u,v,u^{-1},v^{-1}]$ of polynomials in two variables and their inverses over Λ. (Such polynomials in u and u^{-1} are sometimes called "L-polynomials," the L standing for Laurent.)

LEMMA 8.5. *The element* $\{u,v\}$ *generates an infinite cyclic direct summand of the group* $K_2\Lambda[u,v,u^{-1},v^{-1}]$.

The proof can be outlined as follows. Clearly it suffices to consider the case $\Lambda = C$. Let T denote the torus consisting of all pairs (z_1,z_2) of complex numbers with $|z_1| = |z_2| = 1$. Map $C[u,v,u^{-1},v^{-1}]$ to the ring C^T of continuous functions on T by assigning to u and v the continuous functions $u(z_1,z_2) = z_1$ and $v(z_1,z_2) = z_2$ respectively. The proof is now based on the homomorphism

$$\tilde{\phi} : K_2C^T \to \pi_1(SU^T)$$

of §7. The group $\pi_1(SU^T) \cong \pi_1(SU(2)^T)$ is infinite cyclic, and a direct geometric argument shows that ϕ carries the element $\{u,v\}$ of K_2C^T onto a generator of this infinite cyclic group.

I will not try to give further details of this proof, since the method is rather clumsy, and since sharper results can be obtained by using Matsumoto's methods. (Compare §11.4.)

It may be conjectured that the homomorphism

$$K_i C^X \to K^{-i} X$$

of §7, from algebraic K-theory to topological K-theory, is compatible with the product operations in the two theories. If true, this would lead to a quite direct proof of 8.5.

Now let us define the product operation $K_1\Lambda \otimes K_1\Lambda \to K_2\Lambda$. We continue to assume that Λ is commutative.

Given an automorphism α of the free module Λ^m and an automorphism β of the free module Λ^n, we can form the automorphism $\alpha \otimes \beta$ of $\Lambda^m \otimes \Lambda^n$. In order to translate this construction into matrix notation, it is only necessary to choose some fixed ordering for the canonical basis of $\Lambda^m \otimes \Lambda^n$. Then to each matrix A of $GL(m,\Lambda)$ and each matrix B of $GL(n,\Lambda)$ there corresponds the matrix of the corresponding automorphism of $\Lambda^m \otimes \Lambda^n$ which we denote by $A \otimes B \in GL(mn,\Lambda)$. Using this notation, we will construct a bimultiplicative symbol

$$\{A,B\} \in K_2\Lambda$$

which generalizes the symbol $\{u,v\}$.

Let I denote the $m \times m$ identity matrix and I' the $n \times n$ identity matrix. Note that the matrices $A \otimes I'$ and $I \otimes B$ of $GL(mn,\Lambda)$ commute with each other. Hence the matrices

$$\mathrm{diag}(A \otimes I', A^{-1} \otimes I', I \otimes I')$$

and

$$\mathrm{diag}(I \otimes B, I \otimes I', I \otimes B^{-1})$$

of $E(3mn,\Lambda)$ also commute with each other. *Define* $\{A,B\}$ *to be the element*

$$\mathrm{diag}(A \otimes I', A^{-1} \otimes I', I \otimes I') \star \mathrm{diag}(I \otimes B, I \otimes I', I \otimes B^{-1})$$

of $K_2\Lambda$.

Notice that this symbol $\{A,B\}$, for the special case $m = n = 1$, coincides with the symbol $\{u,v\}$.

LEMMA 8.6. *This symbol* $\{A,B\}$ *is well defined, bimultiplicative, and skew-symmetric. Furthermore, if*

$$B = \text{diag}(B_1, B_2)$$

with $B_i \in GL(n_i,\Lambda)$, *then the symbol* $\{A,B\}$ *is equal to* $\{A,B_1\}\{A,B_2\}$.

Proof. The only choice involved in the definition of $\{A,B\}$ was the ordering of the basis elements of $\Lambda^m \otimes \Lambda^n$. If we change this ordering, the result will be to conjugate the relevant matrices in $E(3mn,\Lambda)$ by an appropriate permutation matrix P. Passing to the larger group $E(3mn+1,\Lambda)$, this is the same as conjugating by a matrix $\text{diag}(P,\pm 1)$ which belongs to $E(3mn+1,\Lambda)$. Using Lemma 8.1, it follows that $\{A,B\}$ is well defined.

The proof of skew-symmetry is similar.

In order to prove the bimultiplicative property we will need the following.

LEMMA 8.7. *If the matrices* $X,U \in E(r,\Lambda)$ *commute with each other, and if the matrices* $X',Y' \in E(r',\Lambda)$ *commute with each other, then the symbol*

$$\text{diag}(X,X') \star \text{diag}(Y,Y') \in K_2\Lambda$$

is equal to the product of $X \star Y$ *and* $X' \star Y'$.

This is proved easily, making use of the fact that every generator x_{ij}^{λ} of the Steinberg group with $i,j \le r$ commutes with every $x_{k\ell}^{\mu}$ for which $k,\ell > r$. ∎

Now suppose that the matrix B is equal to B_1B_2. Then the matrix $\text{diag}(I\otimes B, I\otimes I', I\otimes B^{-1})$ is equal to the product of the three matrices $\text{diag}(I\otimes B_1, I\otimes I', I\otimes B_1^{-1})$, $\text{diag}(I\otimes B_2, I\otimes I', I\otimes B_2^{-1})$, and $\text{diag}(I\otimes I', I\otimes I', I\otimes [B_2,B_1$. Hence, using 8.1, the symbol $\{A,B\}$ is equal to the product of the three symbols $\{A,B_1\}$ and $\{A,B_2\}$ and

$$\text{diag}(A\otimes I', A^{-1}\otimes I', I\otimes I') \star \text{diag}(I\otimes I', I\otimes I', I\otimes [B_2,B_1]).$$

But the third factor is equal to 1. For, according to 8.7, after passing to the larger group $E(4mn,\Lambda)$, we can express it as the product of

$$\mathrm{diag}(A \otimes I, A^{-1} \otimes I) \star \mathrm{diag}(I \otimes I', I \otimes I') = 1$$

and

$$\mathrm{diag}(I \otimes I', I \otimes I') \star \mathrm{diag}(I \otimes [B_2, B_1], I \otimes I') = 1.$$

This proves the bimultiplicative property.

Since the formula for $\{A, \mathrm{diag}(B_1, B_2)\}$ is easily verified, using 8.7, this completes the proof of Lemma 8.6. ∎

THEOREM 8.8. *This symbol* $\{A,B\}$ *gives rise to a well defined, skew-symmetric, bimultiplicative pairing from* $K_1\Lambda \times K_1\Lambda$ *to* $K_2\Lambda$.

The proof is immediate. For the rule

$$\begin{aligned}\{A, \mathrm{diag}(B,1)\} &= \{A,B\}\{A,1\} \\ &= \{A,B\},\end{aligned}$$

together with the corresponding rule for A, tells us that the symbol $\{A,B\}$ gives rise to a bimultiplicative function

$$GL(\Lambda) \times GL(\Lambda) \to K_2\Lambda;$$

and since the target group is abelian, we can abelianize the groups on the left.

If we switch to an additive notation for $K_1\Lambda$ and $K_2\Lambda$ then the notation $K_1\Lambda \otimes K_1\Lambda \to K_2\Lambda$ is appropriate.

Just as in §4, this product operation gives rise to a product

$$K_1\Lambda \otimes K_1\mathfrak{a} \to K_2\mathfrak{a}$$

for any two-sided ideal $\mathfrak{a} \subset \Lambda$.

Note that the product $K_1\Lambda \otimes K_1\Lambda \to K_2\Lambda$ is not always surjective:

Example. Let X be the sphere S^2, and consider the ring of continuous real valued functions R^X. It follows from §7 that K_1R^X is equal to the group of units $(R^\bullet)^X$, which splits as the direct sum of a cyclic group of order 2 and an infinitely divisible group. On the other hand K_2R^X has an infinite cyclic direct summand by 7.7. Clearly then the homomorphism

$$K_1R^X \otimes K_1R^X \to K_2R^X$$

cannot be surjective.

To conclude this section, we note that the various product operations we have defined behave well with respect to each other.

LEMMA 8.9. *The associative law holds for the product operation*

$$K_i\Lambda \otimes K_j\Lambda \otimes K_k\Lambda \to K_{i+j+k}\Lambda,$$

$i+j+k \leq 2$. *In particular the product* $K_1\Lambda \otimes K_1\Lambda \to K_2\Lambda$ *is bilinear with respect to the action of* $K_0\Lambda$ *on* $K_1\Lambda$ *and* $K_2\Lambda$.

The proof will be left to the reader. The key step is to note that the homomorphism

$$h_P : GL(n,\Lambda) \to GL(\Lambda),$$

which is used to define the action of $K_0\Lambda$ on $K_1\Lambda$ and $K_2\Lambda$, satisfies the identity

$$(h_P A) \star (h_P B) = h_{P*}(A \star B)$$

for commuting matrices $A, B \in E(n,\Lambda)$. (Compare p. 51.)

§9. Computations in the Steinberg Group

The results in this section are all due to Steinberg. We will work in $St(n,\Lambda)$, $n \geq 3$.

For any unit u of Λ consider the elements

$$w_{ij}(u) = x_{ij}^{u} x_{ji}^{-u^{-1}} x_{ij}^{u} \quad \text{and} \quad h_{ij}(u) = w_{ij}(u)\, w_{ij}(-1).$$

(Note the identities $w_{ij}(u)w_{ij}(-u) = 1$ and $h_{ij}(1) = 1$.) Let $W \subset St(n,\Lambda)$ denote the subgroup generated by all the $w_{ij}(u)$.

A matrix in $GL(n,\Lambda)$ is called a *monomial matrix* if it can be expressed as the product PD of a permutation matrix and a diagonal matrix.

LEMMA 9.1. *If* Λ *is commutative then the image* $\phi(W) \subset GL(n,\Lambda)$ *can be described as the set of all monomial matrices with determinant* 1.

Proof. Inspection shows that $\phi w_{ij}(u)$ is the monomial matrix with entry u in the (i,j)-th place and $-u^{-1}$ in the (j,i)-th place, and that $\phi h_{ij}(u)$ is the diagonal matrix with entry u in the (i,i)-th place and u^{-1} in the (j,j)-th place. These matrices clearly generate the indicated group of monomial matrices. ∎

LEMMA 9.2. *Conjugation by an element of* W *carries each generator* x_{ij}^{λ} *of* $St(n,\Lambda)$ *into some generator* $x_{k\ell}^{\mu}$ *of* $St(n,\Lambda)$.

Before giving the proof of 9.2 let us derive two consequences.

COROLLARY 9.3. *The kernel* C_n *of the homomorphism*

$$\phi|W : W \to GL(n,\Lambda)$$

lies in the center of $St(n,\Lambda)$.

For if $\phi(w) = I$ with $w \in W$, then applying ϕ to the formula

$$w x_{ij}^{\lambda} w^{-1} = x_{k\ell}^{\mu}$$

we obtain

$$I \cdot e_{ij}^{\lambda} \cdot I^{-1} = e_{k\ell}^{\mu}$$

in $GL(n,\Lambda)$; which clearly implies that $x_{ij}^{\lambda} = x_{k\ell}^{\mu}$. ∎

The next corollary is simply a more precise statement of 9.2. Given $w \in W$ express $\phi(w)$ as a product PD, where P is the permutation matrix corresponding to some permutation π of the integers between 1 and n, and D is the diagonal matrix $\mathrm{diag}(v_1, v_2, \ldots, v_n)$.

COROLLARY 9.4. *The conjugate* $w x_{ij}^{\lambda} w^{-1}$ *is equal to* $x_{\pi(ij)}^{v_i \lambda v_j^{-1}}$. *Similarly conjugation by* w *carries* $w_{ij}(u)$ *to* $w_{\pi(ij)}(v_i u v_j^{-1})$ *and* $h_{ij}(u)$ *to* $h_{\pi(ij)}(v_i u v_j^{-1}) h_{\pi(ij)}(v_i v_j^{-1})^{-1}$.

Here $\pi(ij)$ is an abbreviation for the pair $\pi(i)$, $\pi(j)$.

Proof. The first formula can be established by noting that

$$PD\, e_{ij}^{\lambda}\, D^{-1} P^{-1} = e_{\pi(ij)}^{v_i \lambda v_j^{-1}}$$

and then applying 9.2. The other two formulas follow immediately. ∎

Proof of Lemma 9.2. Clearly it suffices to compute $w x_{ij}^{\lambda} w^{-1}$ in case w is a generator $w_{k\ell}(u)$ of W. The proof will be divided into seven cases, depending on which of the two integers i and j are equal to which of k and ℓ. For example if $i = k$ but $j \neq \ell$ then, setting $uv = 1$, we have

$$\begin{aligned} w_{i\ell}(u) x_{ij}^{\lambda} w_{i\ell}(-u) &= x_{i\ell}^{u} x_{\ell i}^{-v} (x_{i\ell}^{u} x_{ij}^{\lambda} x_{i\ell}^{-u}) x_{\ell i}^{v} x_{i\ell}^{-u} \\ &= x_{i\ell}^{u} (x_{\ell i}^{-v} x_{ij}^{\lambda} x_{\ell i}^{v}) x_{i\ell}^{-u} = x_{i\ell}^{u} x_{\ell j}^{-v\lambda} (x_{ij}^{\lambda} x_{i\ell}^{-u}) \\ &= (x_{i\ell}^{u} x_{\ell j}^{-v\lambda} x_{i\ell}^{-u}) x_{ij}^{\lambda} = (x_{ij}^{-\lambda} x_{\ell j}^{-v\lambda}) x_{ij}^{\lambda} = x_{\ell j}^{-v\lambda}. \end{aligned}$$

Similar computations hold whenever at least three of the four indices i,j,k,ℓ are distinct. Finally, if $i = k$ and $j = \ell$, then

$$w_{ij}(u)\,x_{ij}^{\lambda}\,w_{ij}(-u) = w_{ij}(u)\,[x_{iq}^{\lambda},x_{qj}^{1}]\,w_{ij}(-u)$$
$$= [x_{jq}^{-v\lambda},x_{qi}^{v}] = x_{ji}^{-v\lambda v};$$

and similarly

$$w_{ji}(u)\,x_{ij}^{\lambda}\,w_{ji}(-u) = x_{ji}^{-u\lambda u}.$$

This completes the proof. ■

Here are some examples to illustrate 9.4. First let $w = w_{ij}(u)$ so that the associated permutation π transposes i and j, and so that $v_i = -u^{-1}$, $v_j = u$. Conjugating w by itself we obtain

$$w\,w_{ij}(u)\,w^{-1} = w_{ji}(-u^{-1}uu^{-1})$$

or in other words:

LEMMA 9.5. $w_{ij}(u) = w_{ji}(-u^{-1})$.

For the next example, let $w = h_{12}(u)$ so that π is the identity permutation and $v_1 = u$, $v_2 = u^{-1}$. Then

$$w\,h_{13}(v)\,w^{-1} = w(w_{13}(v)\,w_{13}(-1))\,w^{-1}$$

is equal to

$$w_{13}(uv)\,w_{13}(-u) = h_{13}(uv)\,h_{13}(u)^{-1}.$$

Multiplying on the right by $h_{13}(v)^{-1}$, this proves the following.

LEMMA 9.6. *The commutator* $[h_{12}(u),h_{13}(v)]$ *of* §8.2 *is equal to the expression*

$$h_{13}(uv)\,h_{13}(u)^{-1}h_{13}(v)^{-1}.$$

In the commutative case, this expression measures the extent to which the correspondence $u \mapsto h_{13}(u)$ is not a homomorphism.

If Λ is commutative, then clearly these expressions belong to the central group $C_n = W \cap \text{kernel}(\phi)$. Just as in §8 we can then prove:

LEMMA 9.7. *If* Λ *is commutative, then the symbol*

$$\{u,v\} = [h_{ij}(u), h_{ik}(v)]$$
$$= h_{ik}(uv)\, h_{ik}(u)^{-1} h_{ik}(v)^{-1}$$

is skew-symmetric and bi-multiplicative, with values in the commutative group C_n.

No confusion will arise in thus giving the symbol $\{u,v\}$ a new and sharper meaning, since this new symbol clearly corresponds to the old one under the natural homomorphism

$$C_n \to \varinjlim C_n \subset K_2\Lambda.$$

Clearly this symbol $\{u,v\}$ does not depend on the choice of indices $i \neq j \neq k \neq i$.

Here is a fundamental lemma which relates the symbol $\{u,v\}$ to the additive structure of the ring Λ.

LEMMA 9.8. *If both* u *and* $1-u$ *are units, then* $\{u, 1-u\} = 1$. *Furthermore, for any unit* u *the identity* $\{u, -u\} = 1$ *is valid.*

Proof. If v is equal to either $1-u$ or $-u$, we must verify the equality

$$(1) \qquad h_{12}(u) h_{12}(v) = h_{12}(uv).$$

Substituting the definition $h_{12}(x) = w_{12}(x)\, w_{12}(-1)$ into (1), it suffices to prove that the expression

$$(2) \qquad w_{12}(u)\, w_{12}(-1)\, w_{12}(v)$$

is equal to $w_{12}(uv)$.

If $v = 1-u$, this can be proved by substituting the identity

$$w_{12}(-1) = w_{21}(1) = x_{21}^{1} x_{12}^{-1} x_{21}^{1}$$

into (2). (Compare 9.5.) Using the formulas

$$w_{12}(u)\, x_{21}^{1} = x_{12}^{-u^2}\, w_{12}(u)$$

and

$$x_{21}^{1} w_{12}(v) = w_{12}(v) x_{12}^{-v^2}$$

of §9.4, the expression (2) becomes

$$(2') \qquad x_{12}^{-u^2} w_{12}(u) x_{12}^{-1} w_{12}(v) x_{12}^{-v^2} .$$

Now substituting the definitions of $w_{12}(u)$ and $w_{12}(v)$, and using the equalities $-u^2+u = uv$, $u-1+v = 0$, $v-v^2 = uv$, and then $-u^{-1}-v^{-1} = -(uv)^{-1}$, the expression (2′) becomes

$$x_{12}^{uv} x_{21}^{-u^{-1}} x_{12}^{0} x_{21}^{-v^{-1}} x_{12}^{uv} = x_{12}^{uv} x_{21}^{-(uv)^{-1}} x_{12}^{uv} = w_{12}(uv),$$

as required.

The proof that

$$w_{12}(u) w_{12}(-1) w_{12}(-u) = w_{12}(-u^2)$$

is easier. Writing the left hand side as $w_{12}(u) w_{12}(-1) w_{12}(u)^{-1}$, it is equal to $w_{21}(u^{-2})$ by 9.4, and hence to $w_{12}(-u^2)$ by 9.5. This completes the proof of Lemma 9.8. (I want to thank Steinberg for this version of the proof.) ∎

COROLLARY 9.9. *If* Λ *is a finite field, or if* Λ *is the ring of integers modulo a power of an odd prime,** *then* $\{u,v\} = 1$ *for all* u *and* v.

Proof. First suppose that Λ is a field with q elements, q being odd. Then the group $\Lambda^\bullet$ of units is cyclic of order $q-1$. Half of its elements are squares, and half are non-squares. Note that we can find a non-square u_0 so that $1-u_0$ is also a non-square. For otherwise the correspondence $u \mapsto 1-u$ from $\Lambda^\bullet - \{1\}$ to itself would carry each of the $(q-1)/2$ non-squares to a square. Thus we would find $(q-1)/2$ squares in $\Lambda^\bullet$, in addition to 1; which is impossible.

---

* Compare p. 92.

Let v be a generator of the group $\Lambda^\bullet$. Then $u_0 = v^i$, $1-u_0 = v^j$ with i and j odd, and it follows that $\{v,v\}^{ij} = 1$. Since $\{v,v\}^2 = 1$ by skew symmetry, this proves that $\{v,v\}$ itself is trivial.

If $\Lambda = Z/p^nZ$ with p odd, then a similar argument applies since $\Lambda^\bullet$ is cyclic, and since an integer is a quadratic residue modulo p^n if and only if it is a quadratic residue modulo p. Finally, if Λ is a field of order $q = 2^k$, then the equations $\{v,v\}^{q-1} = 1$ and $\{v,v\}^2 = 1$ imply that $\{v,v\} = 1$. This proves the corollary. ■

Next we will show that the group C_n is generated by the symbols $\{u,v\}$. As a first step we will prove the following.

LEMMA 9.10. *All of the symbols* $h_{jk}(u)$ *can be expressed in terms of the* $h_{1k}(u)$ *with* $j = 1$. *In fact these symbols satisfy the identities*

$$h_{jk}(u)\,h_{kj}(u) = 1, \quad \textit{and} \quad h_{ij}(u)^{-1}h_{jk}(u)^{-1}h_{ki}(u)^{-1} = 1.$$

Proof. Using 9.4 to simplify its first three factors, we see that the commutator $h_{ik}(u)\,w_{jk}(1)\,h_{ik}(u)^{-1}w_{jk}(-1)$ is equal to $w_{jk}(u)\,w_{jk}(-1) = h_{jk}(u)$. On the other hand, using 9.4 to simplify its last three factors, this commutator is equal to $h_{ik}(u)\,h_{ij}(u)^{-1}$. Thus:

$$(*) \qquad h_{jk}(u) = h_{ik}(u)\,h_{ij}(u)^{-1}.$$

Setting $i = 1$, we have proved the first assertion of 9.10 (at least in the case $j,k > 1$). Multiplying the identity (*) by the corresponding identity with j and k interchanged, we obtain the second assertion of 9.10, and substituting $h_{ki}(u)^{-1}$ for $h_{ik}(u)$ we obtain the third. ■

REMARK. Evidently the symbols $h_{ij}(u)^{-1}$ would have been somewhat easier to work with than the $h_{ij}(u)$ themselves.

Let Λ be any commutative ring.

THEOREM 9.11. *The central subgroup*

$$C_n = W \cap (\text{kernel } \phi) \subset St(n,\Lambda)$$

is generated by the symbols $\{u,v\}$.

Proof. Let $H \subset W$ denote the subgroup generated by all of the symbols $h_{ij}(u)$. This is a normal subgroup of W by 9.4. We will first show that $C_n \subset H$.

Modulo H we have the relations

$$w_{ij}(u) \equiv w_{ij}(1).$$

Denoting this common residue class by w_{ij}, note that $w_{ij} \equiv w_{ji}$ mod H. Now given any element

$$c = w_{i_1 j_1}(u_1) \dots w_{i_k j_k}(u_k)$$

of C_n, use the relations

$$w_{ij} w_{1\ell} \equiv w_{\pi(1\ell)} w_{ij} \pmod{H}$$

for $i,j > 1$ to push all occurrences of $w_{1\ell}$ to the left. (Here π denotes the permutation which interchanges i and j.) Then use the relations

$$w_{1\ell} w_{1\ell} \equiv 1$$

and

$$w_{1j} w_{1\ell} \equiv w_{1\ell} w_{\ell j} \quad \text{for } j \neq \ell$$

to eliminate the $w_{1\ell}$, one or two at a time. At the end there will be at most a single $w_{1\ell}$ at the left. But this single $w_{1\ell}$ cannot occur since otherwise $\phi(c)$ would not correspond to the identity permutation. Similarly we can eliminate all occurrences of $w_{2\ell}$, and so on, continuing inductively until we have proved that $c \equiv 1$ modulo H.

Thus c can be written as a product of symbols $h_{ij}(u)$. Using 9.10 it follows that c can be written as a product of the symbols $h_{1\ell}(u)$ and their inverses.

If C^* denotes the subgroup of C_n generated by the $\{u,v\}$, note that

$$h_{1\ell}(uv) \equiv h_{1\ell}(u) h_{1\ell}(v) \mod C^*$$

and

$$h_{1j}(u) h_{1\ell}(v) \equiv h_{1\ell}(v) h_{1j}(u) \mod C^*.$$

It follows easily that the element c can be expressed as a product of the form

$$c \equiv h_{12}(u_2)h_{13}(u_3) \ldots h_{1n}(u_n) \mod C^*.$$

This implies that $\phi(c)$ is the diagonal matrix $\operatorname{diag}(u_2 \ldots u_n, u_2^{-1}, \ldots, u_n^{-1})$. But $\phi(c) = I$ by hypothesis, hence $u_2 = u_3 = \ldots = u_n = 1$, and therefore $c \equiv 1 \mod C^*$. This completes the proof of 9.11. ∎

THEOREM 9.12. *If Λ is a field or skew-field, then the entire kernel of*

$$\phi : \mathrm{St}(n,\Lambda) \to \mathrm{GL}(n,\Lambda)$$

is contained in W, and hence is equal to C_n. Therefore the Steinberg group $\mathrm{St}(n,\Lambda)$ is a central extension of $E(n,\Lambda)$ for every $n \geq 3$.

Combining 9.11, 9.12, and 9.9 and passing to the direct limit as $n \to \infty$, this implies the following.

COROLLARY 9.13. *If Λ is a field then $K_2\Lambda$ is generated by the symbols $\{u,v\}$. In particular if Λ is a finite field then $K_2\Lambda = 1$.*

REMARK. M. Stein has recently generalized this result by showing that $K_2\Lambda$ is generated by the symbols $\{u,v\}$ for any semi-local ring Λ which is additively generated by the set $\Lambda^\bullet$ of units.

The proof of 9.12 will be based on two lemmas.

Let T denote the subgroup of $\mathrm{St}(n,\Lambda)$ generated by all x_{ij}^{λ} for which $i < j$.

LEMMA 9.14. *Every element of the group T can be written as a product*

$$\prod_{i<j} x_{ij}^{\lambda(ij)},$$

the factors being arranged in lexicographical order. Hence ϕ maps T isomorphically onto the nilpotent group consisting of all upper triangular matrices with 1's along the diagonal.

The proof is not difficult.

LEMMA 9.15. *If* Λ *is a field or skew-field then every element of* $\mathrm{St}(n,\Lambda)$ *belongs to the product* TWT.

Proof. Clearly $\mathrm{St}(n,\Lambda)$ is generated by those elements x_{ij}^{λ} for which $j = i \pm 1$. Since

$$w_{ij}(1)\, x_{ij}^{\lambda} w_{ij}(-1) = x_{ji}^{-\lambda},$$

it follows that $\mathrm{St}(n,\Lambda)$ is generated by T and the $w_{i,i+1}(1)$. Thus to prove 9.15 it suffices to check that TWT is closed under right multiplication by $w_{i,i+1}(1)$.

To simplify the notation, set $j = i+1$. Given any $t_1 w t_2$ in TWT, write t_2 as a product $x_{ij}^{\lambda} t'$ where t' can be written as a product of $x_{k\ell}^{\mu}$ with $k < \ell$ and $(k,\ell) \neq (i,j)$. Then

$$t_1 w t_2\, w_{ij}(1) = t_1 (w x_{ij}^{\lambda}\, w_{ij}(1)) t'',$$

where the element

$$t'' = w_{ij}(-1)\, t'\, w_{ij}(1)$$

clearly belongs to T. Thus it suffices to check that

$$w x_{ij}^{\lambda}\, w_{ij}(1) \in \mathrm{TWT}.$$

Let π be the permutation corresponding to w.

Case 1. If

$$\pi(i) < \pi(j) = \pi(i+1),$$

then

$$w x_{ij}^{\lambda}\, w_{ij}(1) = x_{\pi(ij)}^{\lambda'}\, w w_{ij}(1) \in \mathrm{TW}.$$

Case 2. If $\pi(i) > \pi(j)$ and λ is a unit, then

$$x_{ij}^{\lambda} = w_{ij}(\lambda)\, x_{ij}^{-\lambda}\, x_{ji}^{\lambda^{-1}}$$

hence

$$\begin{aligned} w x_{ij}^{\lambda}\, w_{ij}(1) &= (w w_{ij}(\lambda)\, x_{ij}^{-\lambda})\, (x_{ji}^{\lambda^{-1}}\, w_{ij}(1)) \\ &= (x_{\pi(ji)}^{\lambda'}\, w w_{ij}(\lambda))\, (w_{ij}(1)\, x_{ij}^{-\lambda^{-1}}) \in \mathrm{TWT}. \end{aligned}$$

Case 3. If λ is not a unit, then $\lambda = 0$ and the statement is clear.

This proves Lemma 9.15. ∎

Proof of Theorem 9.12. If $\phi(t_1 w t_2) = I$ then the monomial matrix $\phi(w)$ is equal to the upper triangular matrix $\phi(t_1)^{-1}\,\phi(t_2)^{-1}$. Hence

$$\phi(w) = \phi(t_1^{-1}\,t_2^{-1}) = I$$

which implies that $w \in C_n$ and $t_2 = t_1^{-1}$ so that

$$t_1 w t_2 = t_1 w t_1^{-1} = w \in C_n$$

as required. This completes the proof. ∎

We will continue the discussion of K_2 of a field in §11.

§10. Computation of K_2Z

This section will prove the following.

THEOREM 10.1. *For* $n \geq 3$ *the group* $St(n,Z)$ *is a central extension*

$$1 \to C_n \to St(n,Z) \to E(n,Z) \to 1$$

where C_n *is the cyclic group of order* 2 *generated by the symbol*

$$\{-1, -1\} = (x_{12}^1 x_{21}^{-1} x_{12}^1)^4.$$

(Compare §§ 9.3, 9.7, and 9.11.) An immediate consequence is the following.

COROLLARY 10.2. *The group* K_2Z *is cyclic of order* 2, *generated by* $\{-1, -1\}$.

For K_2Z is the direct limit of these groups C_n.

The following statement is equivalent to 10.1. Let $e_{ij} = \phi(x_{ij}^1)$ be the elementary matrix with entry $+1$ in the (i, j)-th place, $i \neq j$.

COROLLARY 10.3. *For* $n \geq 3$ *the group* $SL(n, Z) = E(n, Z)$ *has a presentation with* $n(n-1)$ *generators* e_{ij} *subject only to the Steinberg relations*

$$[e_{ij}, e_{k\ell}] = 1 \quad \textit{if } j \neq k,\ i \neq \ell,$$

$$[e_{ij}, e_{jk}] = e_{ik} \quad \textit{if } i, j, k \textit{ are distinct,}$$

and to the relation $(e_{12}e_{21}^{-1}e_{12})^4 = 1$.

Proof of 10.3. Defining e_{ij}^{λ} to be the λ-th power of e_{ij}, all of the Steinberg relations between these elements e_{ij}^{λ} follow immediately from the relations listed. Together with 10.1, this completes the proof. ■

The proof of 10.1 is based on classical work by Nielsen and Magnus. In fact 10.1 can be derived from Magnus' paper *Über n-dimensionale Gittertransformationen,* Acta Math. 64 (1935), 353-367. We will give a simplified proof, based on recent work by J. R. Silvester. (See Silvester, *A Presentation of* $GL_n(Z)$ *and* $GL_n(k[X])$, to appear.) It should be noted that Silvester, and also K. Dennis, have used this same method to prove the isomorphism $K_2F[x] \cong K_2F$, where $F[x]$ denotes the ring of polynomials in one indeterminate over the field F. Dennis has extended this to the case of a skew-field.

Since Silvester's proof is inductive, we must start with the cases $n = 1, 2$. For $n = 1$, we simply define $St(1,\Lambda)$ to be trivial. For $n = 2$, the definition according to Steinberg is as follows. (Compare p. 73.)

DEFINITION 10.4. For any ring Λ (associative with 1) let $St(2,\Lambda)$ be the group with generators x_{12}^{λ} and x_{21}^{λ}, as λ varies over Λ, and defining relations

$$x_{ij}^{\lambda}x_{ij}^{\mu} = x_{ij}^{\lambda+\mu}$$

and

$$w_{ij}(u)\,x_{ji}^{\lambda}\,w_{ij}(-u) = x_{ij}^{-u\lambda u}$$

for $u \in \Lambda^{\bullet}$; where $w_{ij}(u)$ by definition is equal to $x_{ij}^{u}x_{ji}^{-u^{-1}}x_{ij}^{u}$.

The analogue of 10.1 is then the following. We introduce the abbreviation x_{ij} for the Steinberg generator x_{ij}^{1}.

THEOREM 10.5. *The group* $St(2,Z)$ *is a central extension*

$$1 \to C_2 \to St(2,Z) \to E(2,Z) \to 1$$

where C_2 *is a free cyclic group generated by the element* $(x_{12}x_{21}^{-1}x_{12})^4$. *Hence* $SL(2,Z) = E(2,Z)$ *has presentation with generators* e_{12} *and* e_{21} *and defining relations*

$$e_{12}e_{21}^{-1}e_{12} = e_{21}^{-1}e_{12}e_{21}^{-1} \quad \textit{and} \quad (e_{12}e_{21}^{-1}e_{12})^4 = 1.$$

The second statement (which is of course classical) follows from the first, since inspection shows that the two generators x_{12} and x_{21} of St(2,Z) are subject only to the relation

$$x_{12}x_{21}^{-1}x_{12} = x_{21}^{-1}x_{12}x_{21}^{-1}.$$

Miscellaneous remarks. The group with two generators $\alpha = x_{12}$ and $\beta = x_{21}^{-1}$ and one defining relation $\alpha\beta\alpha = \beta\alpha\beta$ is familiar in knot theory as the fundamental group of the complement of a trefoil knot $K \subset S^3$. (This group can also be presented by generators ζ, ω with defining relation $\zeta^3 = \omega^2$; where $\zeta = \alpha\beta$ and $\omega = \alpha\beta\alpha$.)

This rather curious isomorphism $St(2,Z) \cong \pi_1(S^3 - K)$ does have a geometric explanation. Think of $E(2,Z) = SL(2,Z)$ as a discrete subgroup of the connected Lie group $E(2,R) = SL(2,R)$. The universal covering $\tilde{E}(2,R)$ is a central extension

$$1 \to \Pi \to \tilde{E}(2,R) \to E(2,R) \to 1,$$

where Π denotes the fundamental group

$$\Pi = \pi_1 E(2,R) \cong Z.$$

Define $\tilde{E}(2,Z)$ to be the inverse image of the subgroup $E(2,Z) \subset E(2,R)$ in $\tilde{E}(2,R)$. Then evidently the sequence

$$1 \to \Pi \to \tilde{E}(2,Z) \to E(2,Z) \to 1$$

is exact, and Π is central in $\tilde{E}(2,Z)$. It is not difficult to show that this last sequence can be identified with the sequence of 10.5; so that $\tilde{E}(2,Z) \cong St(2,Z)$.

This group $\tilde{E}(2,Z)$ can also be identified with the fundamental group of the homogeneous space

$$M^3 = E(2,R)/E(2,Z).$$

In fact the composition of the two covering maps

$$\tilde{E}(2,R) \to E(2,R) \to M^3$$

is precisely the universal covering of M^3, and the group of covering transformations is just the subgroup $\tilde{E}(2,Z)$, acting on $\tilde{E}(2,R)$ by left translation.

ASSERTION. *The homogeneous space* $M^3 = SL(2,R)/SL(2,Z)$ *is diffeomorphic to the complement of a trefoil knot in the sphere* S^3.

The following ingenious proof is due to D. Quillen. First note that M^3 can be identified with the set consisting of all lattices* L in the plane R^2 such that the quotient R^2/L has unit area. For the group $SL(2,R)$ permutes the set of all such lattices transitively, and the subgroup carrying the standard lattice Z^2 to itself is precisely $SL(2,Z)$.

Now replace R^2 by the plane C of complex numbers. To any lattice $L \subset C$ one associates the Weierstrass function $\wp(z)$, a doubly periodic meromorphic function with period lattice L, and with poles of the form

$$\wp(\lambda + z) = z^{-2} + a_2 z^2 + a_4 z^4 + \ldots$$

precisely at the lattice points λ. (See for example Copson, *Functions of a Complex Variable*.) This function $\wp(z)$ satisfies a differential equation of the form

$$(d\wp/dz)^2 = 4\wp^3 - g_2\wp - g_3$$

where the complex constants g_2 and g_3 can be defined by the formulas

$$g_2 = 60 \sum \lambda^{-4}, \quad g_3 = 140 \sum \lambda^{-6},$$

to be summed over all lattice points $\lambda \neq 0$. These constants g_2 and g_3 characterize the function $\wp$ and the lattice L uniquely. A given pair (g_2, g_3) can occur if and only if $27(g_3)^2 - (g_2)^3 \neq 0$. (This expression is the discriminant of the polynomial $4w^3 - g_2 w - g_3$. Its non-vanishing is equivalent to the geometric requirement that the mapping $C/L \to C \cup \infty$ of degree 2 induced by $\wp$ must have four distinct ramification points.) Thus the space $GL(2,R)/GL(2,Z)$ consisting of all lattices in C is diffeomorphic to the complement of the locus

$$27(g_3)^2 - (g_2)^3 = 0$$

---

* Here a lattice means an additive subgroup spanned by two vectors which are linearly independent over R.

in C^2. Let K be the trefoil knot obtained by intersecting this locus with the unit sphere S^3 in C^2. The required diffeomorphism $M^3 \to S^3 - K$ is now constructed as follows. For each lattice L representing an element of M^3 there is a unique "expanded" lattice tL, with $t > 0$, whose invariants $(t^{-4}g_2, t^{-6}g_3)$ lie on the unit sphere. The correspondence $L \mapsto (t^{-4}g_2, t^{-6}g_3)$ defines the required diffeomorphism. ∎

Now let us begin the proofs of Theorems 10.1 and 10.5. Let the group St(n,Z) operate on the right of the module Z^n, consisting of all n-tuples of integers, by means of the natural homomorphism $St(n,Z) \to E(n,Z)$. As an example, for $n = 2$ the action is determined by the formulas

$$(a,b)x_{12} = (a, a+b), \qquad (a,b)x_{21} = (a+b, b).$$

Define a "norm" on Z^n by

$$\|(a_1,\ldots,a_n)\| = |a_1| + \ldots + |a_n|.$$

Let $\pm\beta$ be one of the standard basis vectors of Z^n, so that $\|\beta\| = 1$.

As in §9, the subgroup of St(n,Z) generated by the elements

$$w_{ij} = w_{ij}(1) = x_{ij}x_{ji}^{-1}x_{ij}$$

will be denoted by $W = W_n$. Note that the action of W on Z^n preserves the norm.

LEMMA 10.6 (Silvester). *Every element in* St(n,Z), $n \geq 2$, *can be expressed as a product*

$$g_1 g_2 \cdots g_r w$$

with $w \in W$, *where each* g_α *is one of the Steinberg generators* $x_{ij}^{\pm 1}$, *in such a way that*

$$\|\beta g_1\| \leq \|\beta g_1 g_2\| \leq \cdots \leq \|\beta g_1 g_2 \cdots g_r\| .$$

Proof. To any sequence $g_1, g_2, \ldots, g_r$ of Steinberg generators we associate the sequence $1 = \sigma_0, \sigma_1, \sigma_2, \ldots, \sigma_r$ of positive integers defined by the formula

$$\sigma_\alpha = \|\beta g_1 g_2 \ldots g_\alpha\| .$$

The deviation of this sequence $1, \sigma_1, \ldots, \sigma_r$ from monotonicity can be roughly measured by a pair (λ, μ) of positive integers, defined as follows. If the sequence is monotone $(1 \leq \sigma_1 \leq \sigma_2 \leq \ldots \leq \sigma_r)$, we set $\lambda = \mu = 1$. Otherwise there exists some α with $\sigma_\alpha > \sigma_{\alpha+1}$. Let

$$\lambda = \operatorname{Max}\{\sigma_\alpha \text{ such that } \sigma_\alpha > \sigma_{\alpha+1}\}$$

be the largest value of σ_α for which this occurs. Since there may be several distinct values of α for which this same maximum is attained, we set

$$\mu = \operatorname{Max}\{\alpha \text{ such that } \sigma_\alpha = \lambda > \sigma_{\alpha+1}\} .$$

The pairs (λ, μ) are to be ordered lexicographically; so that $(\lambda, \mu) > (\lambda', \mu')$ if and only if either $\lambda > \lambda'$, or $\lambda = \lambda'$ and $\mu > \mu'$.

The proof of 10.6 will consist in showing that each word $g_1 \ldots g_r w$ with $(\lambda, \mu) > (1, 1)$ can be altered, by means of Steinberg relations, so as to decrease the pair (λ, μ). After repeating this procedure a finite number of times, we must achieve the required word with $(\lambda, \mu) = (1, 1)$.

Suppose then that the pair $(\lambda, \mu) > (1, 1)$. Then

$$\lambda = \sigma_\mu > \sigma_{\mu+1} .$$

Note that $\mu \geq 1$ (since if $\mu = 0$ we would have $1 = \sigma_0 > \sigma_1$ which is impossible); and that $\sigma_{\mu-1} \leq \sigma_\mu$.

We will suppose, to fix our ideas and simplify the notation, that the generator g_μ is equal to x_{12}. (The general case can easily be reduced to this special case as follows. If $g_\mu = x_{ij}$, we can simply renumber the coordinates to replace i and j by 1 and 2 respectively. If $g_\mu = x_{ij}^{-1}$, then conjugating every g_α by w_{ij}, and replacing β by βw_{ij}^{-1} and w by $w_{ij}w$, we obtain an equivalent problem, with g_μ replaced by x_{ji}. We then proceed as before.)

Assume then that $g_\mu = x_{12}$. Setting

$$\beta g_1 g_2 \ldots g_\mu = (a, b, c, \ldots) \in Z^n ,$$

it follows that

$$\beta g_1 g_2 \cdots g_{\mu-1} = (a, b-a, c, \ldots).$$

Hence the inequality $\sigma_{\mu-1} \leq \sigma_\mu$ can be written as

$$|b - a| \leq |b| .$$

This is clearly equivalent to the following statement.

(1) $|a| \leq 2|b|$, *and if* $a \neq 0$ *then* $ab > 0$.

The proof of 10.6 will be divided into seven cases, depending on $g_{\mu+1}$. First we consider the four cases where $g_{\mu+1}$ commutes with $g_\mu = x_{12}$.

Case 1. Suppose that $g_{\mu+1} = x_{1j}^{\varepsilon}$ with $j \geq 3$. Without loss of generality, we may assume that $j = 3$, and $n = 3$. Thus $g_{\mu+1}$ transforms (a, b, c) to $(a, b, \varepsilon a + c)$, with

$$|c| > |\varepsilon a + c| .$$

Now alter the word $g_1 \ldots g_r w$ by replacing the product $g_\mu g_{\mu+1} = x_{12}x_{13}^{\varepsilon}$ by $x_{13}^{\varepsilon}x_{12}$. Then the transformation

$$(a, b-a, c) \overset{g_\mu}{\longmapsto} (a, b, c) \overset{g_{\mu+1}}{\longmapsto} (a, b, \;\; a + c)$$

will be replaced by

$$(a, b-a, c) \mapsto (a, b-a, \varepsilon a + c) \mapsto (a, b, \varepsilon a + c).$$

The associated numbers σ_α are unchanged, except that $\sigma_\mu = \|(a, b, c)\|$ is replaced by a number $\sigma'_\mu = \|(a, b-a, \varepsilon a + c)\|$ which satisfies

$$\sigma_{\mu-1} > \sigma'_\mu .$$

Inspection shows that the pair (λ', μ') associated with the new sequence is less than (λ, μ).

Case 2. If $g_{\mu+1} = x_{ij}^{\varepsilon}$ with i and j greater than 2, the proof proceeds exactly as in Case 1.

Case 3. Suppose that $g_{\mu+1} = g_\mu^{\varepsilon} = x_{12}^{\varepsilon}$. If $\varepsilon = -1$, then we can simply cancel the factor $g_\mu g_{\mu+1}$, thus reducing (λ, μ). But ε cannot be $+1$, for if

$$(a,\ b,\ c)\,g_{\mu+1} = (a,\ b{+}a,\ c)$$

then we must have

$$|b-a| \le |b| > |b+a|$$

which is clearly impossible.

Case 4. If $g_{\mu+1} = x_{32}^{\varepsilon}$ (or x_{i2}^{ε} with $i \ge 3$) then a rather more complicated argument is needed. Note that the product $x_{12}x_{32}^{\varepsilon}$ in the Steinberg group can also be written either as $x_{32}^{\varepsilon}x_{12}$, or as $x_{13}^{\varepsilon}x_{32}^{\varepsilon}x_{13}^{-\varepsilon}$, or as $x_{31}^{\varepsilon}x_{12}x_{31}^{-\varepsilon}$. (The proof of these identities is straightforward.) This means that the transformation

$$(a,\ b-a,\ c) \mapsto (a,\ b,\ c) \mapsto (a,\ b+\varepsilon c,\ c)$$

corresponding to the product $x_{12}x_{32}^{\varepsilon}$ can be replaced either by

$$(a,\ b-a,\ c) \mapsto (a,\ b-a+\varepsilon c,\ c) \mapsto (a,\ b+\varepsilon c,\ c),$$

or by

$$(a,\ b-a,\ c) \mapsto (a,\ b-a,\ c{+}\varepsilon a) \mapsto (a,\ b{+}\varepsilon c,\ c{+}\varepsilon a) \mapsto (a,\ b{+}\varepsilon c,\ c),$$

or by

$$(a,\ b-a,\ c) \mapsto (a{+}\varepsilon c,\ b-a,\ c) \mapsto (a{+}\varepsilon c,\ b{+}\varepsilon c,\ c) \mapsto (a,\ b{+}\varepsilon c,\ c).$$

Inspection shows that the first replacement will reduce (λ,μ) providing that

(2) $$|b-a| > |b-a+\varepsilon c|\ .$$

Similarly the second will reduce $(\lambda,\ \mu)$ if

(3) $$|c| > |c+\varepsilon a|\ ,$$

and the third will reduce $(\lambda,\ \mu)$ if

(4) $$|a| > |a+\varepsilon c|\ .$$

We must show that at least one of these three inequalities is satisfied. Note first that the inequality $\sigma_{\mu} > \sigma_{\mu+1}$ implies that

$$|b| > |b+\varepsilon c|\ .$$

Hence b and εc must have opposite sign. If $a \ne 0$, then it follows from (1) that a and εc have opposite sign, and therefore that either (3) or (4) is satisfied. But if $a = 0$, then (2) is satisfied. This completes the discussion for Case 4.

Now we must consider the various cases in which g_μ and $g_{\mu+1}$ do not commute.

Case 5. If $g_{\mu+1} = x_{21}^{\varepsilon}$, then the product $g_\mu g_{\mu+1}$ corresponds to the transformation

$$(a,\ b-a) \mapsto (a,\ b) \mapsto (a+\varepsilon b,\ b) \tag{5}$$

with

$$|a| > |a+\varepsilon b| \ .$$

If ε were $+1$, then a and b would have opposite sign, contradicting (1); so we may assume that $\varepsilon = -1$. Replacing the product $x_{12}x_{21}^{-1}$ by $x_{21}w_{21}(-1)$, and noting that the element $w_{21}(-1)$ can be pushed innocuously past $g_{\mu+2}g_{\mu+3} \cdots g_r$ by §9.4 or §10.4, we see that the transformation (5) can be replaced by

$$(a,\ b-a) \overset{x_{21}}{\longmapsto} (b,\ b-a).$$

Evidently this reduces $(\lambda,\ \mu)$.

Case 6. Suppose that $g_{\mu+1}$ equals x_{23}^{ε} (or x_{2j}^{ε} with $j \geq 3$). In this case the word $x_{12}x_{23}^{\varepsilon}$ can be replaced either by $x_{13}^{\varepsilon}x_{23}^{\varepsilon}x_{12}$ or by $x_{23}^{\varepsilon}x_{13}^{\varepsilon}x_{12}$ or by $x_{21}x_{13}^{\varepsilon}x_{12}^{-1}w_{12}$. Correspondingly the transformation

$$(a,\ b-a,\ c) \mapsto (a,\ b,\ c) \mapsto (a,\ b,\ c+\varepsilon b)$$

can be replaced by one of the following:

$$(a,\ b-a,\ c) \mapsto (a,\ b-a,\ c+\varepsilon a) \mapsto (a,\ b-a,\ c+\varepsilon b) \mapsto (a,\ b,\ c+\varepsilon b),$$

$$(a,\ b-a,\ c) \mapsto (a,\ b-a,\ c+\varepsilon b-\varepsilon a) \mapsto (a,\ b-a,\ c+\varepsilon b) \mapsto (a,\ b,\ c+\varepsilon b),$$

or

$$(a,\ b-a,\ c) \mapsto (b,\ b-a,\ c) \mapsto (b,\ b-a,\ c+\varepsilon b) \mapsto (b,\ -a,\ c+\varepsilon b).$$

Thus (λ,μ) can be reduced if either

$$|c| > |c+\varepsilon a| \ , \tag{6}$$

$$|c| > |c+\varepsilon b-\varepsilon a| \ , \text{ or} \tag{7}$$

$$|a| > |b| \ ; \tag{8}$$

using the inequality $|b-a| \leq |b|$. The inequality $\sigma_\mu > \sigma_{\mu+1}$ implies that

$$|c| > |c+\varepsilon b|,$$

so that c and εb have opposite sign, and

$$|b| < 2|c|.$$

If $a = 0$, then the equality (7) is satisfied. But if $a \neq 0$, so that $ab > 0$ by (1), then εa and c have opposite sign. Now either $|a| < 2|c|$, which implies (6), or $|a| \geq 2|c|$ which implies (8).

Case 7. If $g_{\mu+1} = x_{31}^{\varepsilon}$ (or x_{i1}^{ε} with $i \geq 3$), then $x_{12}x_{31}^{\varepsilon}$ is equal to $x_{31}^{\varepsilon}x_{12}x_{32}^{-\varepsilon}$ and to $x_{31}^{\varepsilon}x_{13}^{-\varepsilon}x_{32}^{-\varepsilon}x_{13}^{\varepsilon}$. Hence the transformation

$$(a,\ b-a,\ c) \mapsto (a,\ b,\ c) \mapsto (a+\varepsilon c,\ b,\ c)$$

can be replaced either by

$$(a,\ b-a,\ c) \mapsto (a+\varepsilon c,\ b-a,\ c) \mapsto (a+\varepsilon c,\ b+\varepsilon c,\ c) \mapsto (a+\varepsilon c,\ b,\ c)$$

or by

$$(a, b-a, c) \mapsto (a+\varepsilon c, b-a, c) \mapsto (a+\varepsilon c, b-a, -\varepsilon a) \mapsto (a+\varepsilon c, b, -\varepsilon a) \mapsto (a+\varepsilon c, b, c).$$

Hence (λ, μ) can be reduced if either

(9) $$|b+\varepsilon c| \leq |b|, \text{ or}$$

(10) $$|c| \geq |a|.$$

The inequality $\sigma_{\mu} > \sigma_{\mu+1}$ in this case implies that

$$|a| > |a+\varepsilon c|$$

so that a and εc have opposite sign, and

$$|c| < 2|a|.$$

Therefore b and εc have opposite sign (by (1)), so either $|c| \leq 2|b|$ which implies (9), or $|c| \geq 2|b|$ which together with (1) implies (10). This completes the proof of Silvester's lemma. ∎

The next step in the proof of 10.1 and 10.5 is the following.

LEMMA 10.7. *For* $n \geq 2$ *the kernel of the natural homomorphism* ϕ: $St(n,Z) \to E(n,Z)$ *is contained in the subgroup* W_n.

Proof by induction on n. Since the statement is certainly true for $n = 1$, we may assume $n \geq 2$. As standard vector β in Z^n, take the n-th basis vector $(0, 0, \ldots, 0, 1)$. By Silvester's lemma, any given element in the kernel of ϕ can be written as a product $g_1 \ldots g_r w$ with

$$1 \leq \|\beta g_1\| \leq \|\beta g_1 g_2\| \leq \cdots \leq \|\beta g_1 \ldots g_r w\| = 1.$$

The equation $\|\beta g_1\| = 1$ implies that the Steinberg generator g_1 must leave β fixed, and it follows inductively that every g_α must leave β fixed. Further, since $\phi(g_1 \ldots g_r w) = 1$ the element $w \in W_n$ must also leave β fixed.

Thus the word $g_1 \ldots g_r$ cannot contain any Steinberg generator x_{ij}^{ε} with $i = n$. It may contain some x_{ij}^{ε} with $j = n$, but if so, using the Steinberg relations, we can push all such x_{in}^{ε} to the left. Setting the product of all these x_{in}^{ε} equal to x, this means that we can write $g_1 \ldots g_r w$ as a product

$$x\iota(y)w$$

where ι denotes the natural homomorphism

$$\iota : St(n-1, Z) \to St(n, Z).$$

Taking the case $n = 4$ for illustrative purposes, this means that the two matrices $\phi(x)$ and $\phi(\iota(y)w)$ have the form

$$\begin{pmatrix} 1 & 0 & 0 & * \\ 0 & 1 & 0 & * \\ 0 & 0 & 1 & * \\ 0 & 0 & 0 & 1 \end{pmatrix} \text{ and } \begin{pmatrix} * & * & * & 0 \\ * & * & * & 0 \\ * & * & * & 0 \\ 0 & 0 & 0 & 1 \end{pmatrix}$$

respectively. But the product of these matrices is I, so

$$\phi(x) = \phi(\iota(y)w) = I,$$

and it follows from §5.2 that $x = 1$.

Now write w as a product $\iota(w')c$ with

$$w' \in W_{n-1}, \quad c \in W_n \cap \text{kernel}(\phi).$$

It follows that the original element $g_1 \cdots g_r w$ of kernel(ϕ) can be expressed as a product $\iota(yw')c$. Now yw' belongs to the kernel of

$$\mathrm{St}(n-1, Z) \to E(n-1, Z),$$

hence $yw' \in W_{n-1}$ by the induction hypothesis. Therefore $\iota(yw') \in W_n$, which proves Lemma 10.7. ■

The main theorems now follow easily.

Proof of 10.1. Since kernel(ϕ) is contained in W_n, it follows from 9.3 and 9.11 that this kernel is central, and generated by $\{-1, -1\}$. But according to 9.7 and 8.4, the element $\{-1, -1\} \in \mathrm{St}(n, Z)$ has order precisely equal to 2. ■

Proof of 10.5. The relation $w_{12}w_{21} = 1$ in St(2, Z) implies that the group W_2 is cyclic. Since $\phi(w_{12})$ has order 4, the kernel of ϕ is generated by $(w_{12})^4$. As in §9.3, this kernel is central. The fact that no power of w_{12} is trivial in St(2, Z) can be proved by constructing a homomorphism

$$\tilde{\phi} : \mathrm{St}(2, R) \to \tilde{E}(2, R)$$

as in §7.5. ■

Here is another consequence of 10.1. Given any integer $m > 1$, consider the exact sequence (§6.2):

$$\cdots \to K_2 Z \to K_2(Z/mZ) \to K_1(mZ) \to K_1 Z \to \cdots .$$

Using the Mennicke, Bass, Lazard, Serre theorem that $SK_1(mZ) = 1$, we conclude that the group $K_2(Z/mZ)$ is also generated by $\{-1, -1\}$. If m is a power of an odd prime, it follows from §9.9 that $K_2(Z/mZ) = 1$. More generally, using the identity

$$K_2(\Lambda \times \Lambda') \cong K_2\Lambda \times K_2\Lambda'$$

and the Chinese remainder theorem, this implies the following.

COROLLARY 10.8. *If* $m \not\equiv 0 \bmod 4$ *then* $K_2(Z/mZ) = 1$.

For $m \equiv 0 \bmod 4$, K. Dennis has recently shown that $K_2(Z/mZ)$ has order 2.

§11. Matsumoto's Computation of K_2 of a Field

For any field F, the group K_2F has been described by generators and relations in the thesis of H. Matsumoto, based upon earlier work by C. Moore. (See the list of references in the Preface.)

This section will state Matsumoto's theorem, and then derive some consequences from it. In particular, following a letter from John Tate, the group K_2Q will be determined completely. Here Q denotes the field of rational numbers.

If F is a field, then we have seen in §9 that the group K_2F is generated by certain symbols $\{x,y\}$. Here x and y range over the multiplicative group $F^\bullet = F - \{0\}$.

THEOREM 11.1 (Matsumoto). *The abelian group* K_2F *has a presentation, in terms of generators and relations, as follows. The given generators* $\{x,y\}$, *with* x *and* y *in* $F^\bullet$, *are subject only to the following relations and their consequences:*

(1) $\{x,1-x\} = 1$ *for* $x \neq 0,1$,

(2) $\{x_1x_2,y\} = \{x_1,y\}\{x_2,y\}$,

and (3) $\{x,y_1y_2\} = \{x,y_1\}\{x,y_2\}$.

The proof will be given in §12.

Since the kernel C_n of the central extension $St(n,F) \to SL(n,F)$ is generated by corresponding symbols $\{x,y\}$ which satisfy all of these relations,* there is a canonical surjection $K_2F \to C_n$. The following is an immediate consequence.

---

* See §9.7, 9.8, 9.11, and 9.12.

COROLLARY 11.2. *For any field* F, *the groups*

$$C_3 \to C_4 \to C_5 \to \cdots$$

are canonically isomorphic to each other, and to their direct limit K_2F.

It follows that the Schur multiplier $H_2SL(n,F)$ is also isomorphic to K_2F, for any $n \geq 3$, providing that we exclude the exceptional case $n = 3$ and $|F| = 2$ or 4, and $n = 4$, $|F| = 2$. (Compare p. 48.)

Steinberg Symbols

Here is a reformulation of Theorem 11.1.

COROLLARY 11.3. *Given any bimultiplicative symbol*

$$x,y \mapsto c(x,y)$$

on $F^{\bullet}$, *with values in a multiplicative abelian group* A, *satisfying the identity*

$$c(x,1-x) = 1,$$

there exists one and only one homomorphism from K_2F *to* A *which carries the symbol* $\{x,y\}$ *to* $c(x,y)$ *for all* x *and* y.

Any such bimultiplicative mapping $c : F^{\bullet} \times F^{\bullet} \to A$ to an abelian group, satisfying $c(x,1-x) = 1$, will be called a *Steinberg symbol* on the field F.

A classical example of such a Steinberg symbol is the Hilbert quadratic residue symbol in a local field, which can be defined by setting $c(a,b)$ equal to $+1$ or -1 according as the equation $ax^2 + by^2 = 1$ does or does not have a solution x,y in the field. (See for example O'Meara, *Introduction to Quadratic Forms*, p. 164.) More generally, if F is a local field containing the n-th roots of unity, then the n-th power norm residue symbol, with values in the group of n-th roots of unity, can be defined. (See §15.9.)

Note that a Steinberg symbol is necessarily skew-symmetric

$$c(x,y) = c(y,x)^{-1},$$

and necessarily satisfies the identity

$$c(x,-x) = 1.$$

In fact, since $-x = (1-x)/(1-x^{-1})$, the identity $c(x,-x) = 1$ can be proved by dividing the equation $c(x,1-x) = 1$ by $c(x,1-x^{-1}) = c(x^{-1},1-x^{-1})^{-1} = 1$. Now multiplying $c(x,y)$ by $c(x,-x)$, dividing by $c(xy,-xy)$, and then multiplying by $c(y^{-1},-y^{-1})$, we obtain the equation

$$c(x,y) = c(x,-xy) = c(y^{-1},-xy) = c(y^{-1},x),$$

which proves skew-symmetry. We will make frequent use of these facts in §12.

REMARK. An important consequence of 11.2 and 11.3 is the following. *There is a one-to-one correspondence between Steinberg symbols on* F *with values in* A *and central extensions of* SL(n,F) *with kernel* A. (Here we assume that $n \geq 3$, and exclude the three exceptional cases $SL(3, F_2)$, $SL(3, F_4)$, and $SL(4, F_2)$.) In fact any central extension

$$1 \to A \to G \overset{\psi}{\to} SL(n, F) \to 1$$

determines a Steinberg symbol

$$c(u,v) = \left[\psi^{-1}\begin{pmatrix} u & 0 & 0 \\ 0 & u^{-1} & 0 \\ 0 & 0 & 1 \end{pmatrix}, \psi^{-1}\begin{pmatrix} v & 0 & 0 \\ 0 & 1 & 0 \\ 0 & 0 & v^{-1} \end{pmatrix}\right] = \gamma(\{u,v\}),$$

where γ denotes the unique homomorphism from St(n,F) to G over SL(n,F). (Compare p. 48.) Conversely, given a Steinberg symbol c, let

$$N \subset C_n \times A \subset St(n,F) \times A$$

be the graph of the homomorphism

$$\{u,v\} \mapsto c(u,v)^{-1}$$

from C_n to A. Then

$$G = (St(n,F) \times A)/N$$

is the required central extension, with

$$1 \to A \to G \to SL(n,F) \to 1.$$

Expressed in terms of homological algebra, the collection of all isomorphism classes of central extensions with kernel A forms a group

$$H^2(SL(n,F); A) \cong \mathrm{Hom}(H_2 SL(n,F), A) \cong \mathrm{Hom}(K_2 F, A)$$

which is isomorphic to the group of Steinberg symbols with values in A.

If the field F has a topology, so that SL(n,F) is a topological group, it is natural to look for central extensions which are also topological groups.

DEFINITION. If S is a Hausdorff topological group, then a *topological group extension* will mean an exact sequence

$$1 \to A \overset{\iota}{\to} G \overset{\psi}{\to} S \to 1$$

where A and G are Hausdorff topological groups, the homomorphism ι being continuous and closed, and the homomorphism ψ being continuous and open. (These last conditions are equivalent to the requirement that A embeds homeomorphically as a closed subgroup of G, and that the quotient G/A maps homeomorphically onto S.)

Matsumoto proves the following. We assume that F is Hausdorff, with continuous addition, multiplication, and division, and that A is a commutative Hausdorff topological group.

ASSERTION 11.4. *A Steinberg symbol* c *on* F *with values in* A *gives rise to a topological central extension of* SL(n,F) *if and only if* c *is continuous as a function of two variables and satisfies the condition*

$$\lim_{a,b\to 0} c(a,\ 1+ab) = 1.$$

For the proof that these conditions are sufficient, we refer the reader to Matsumoto. The proof that they are necessary can be sketched as follows. Given such a topological central extension, let γ be the unique homomorphism from St(n,F) to G over SL(n,F). We will first show that the homomorphism

$$b \mapsto \gamma(x_{12}^b)$$

from F to G is continuous. Clearly it suffices to prove continuity as b tends to 0. The function

$$g \mapsto [\gamma(x_{13}^1), g]$$

from G to itself is certainly continuous. Hence, given a neighborhood U of the identity in G, there exists a neighborhood V of the identity so that $[\gamma(x_{13}^1), g] \in U$ whenever $g \in V$. Since the function

$$b \mapsto e_{32}^b = \psi(\gamma(x_{32}^b))$$

from F to SL(n,F) is certainly continuous, and since $\psi(V)$ is an open subset of SL(n,F), there exists a neighborhood N of 0 in F so that $e_{32}^b \in \psi(V)$ whenever $b \in N$. Now for any b in N, choosing $g \in V$ so that $\psi(g) = e_{32}^b$, it follows that

$$[\gamma(x_{13}^1), g] = [\gamma(x_{13}^1), \gamma(x_{32}^b)] = \gamma(x_{12}^b)$$

belongs to U. Thus $\gamma(x_{12}^b)$ depends continuously on b.

It follows immediately that the group elements $\gamma(w_{ij}(u))$ and $\gamma(h_{ij}(u))$ depend continuously on the parameter u. Therefore the symbol

$$c(u,v) = \gamma(\{u,v\}) = \gamma([h_{12}(u), h_{13}(v)])$$

is continuous as a function of two variables.

Matsumoto now invokes the following identity. Suppose that we try to write the element $x_{12}^a x_{21}^b$ of St(n,F) as a product of the form $x_{21}^\lambda w x_{12}^\mu$. Using the methods of §9.15 and a little hard work, we arrive at the formula

$$x_{12}^a x_{21}^b = x_{21}^{bd} h_{21}(d) x_{12}^{ad} \{a, d^{-1}\}$$

with $d = (1+ab)^{-1}$. Thus if a and b tend to 0 in F, then d tends to 1 in F and the group elements $\gamma(x_{12}^a x_{21}^b)$ and $\gamma(x_{21}^{bd} h_{21}(d) x_{12}^{ad})$ tend to the identity in G. Hence the symbol

$$\gamma(\{a, d^{-1}\}) = c(a, d^{-1}) = c(a, 1+ab)$$

must also tend to the identity element. This completes our discussion of 11.4.

Here is one easy method for constructing Steinberg symbols. Recall that a *discrete valuation* v on a field F is a homomorphism from the multiplicative group $F^\bullet$ onto the additive group of integers, satisfying $v(x+y) \geq \mathrm{Min}(v(x), v(y))$. The associated *valuation ring* $\Lambda \subset F$ consists of all x with $v(x) \geq 0$, together with the zero element of F. There is a unique maximal ideal $\mathfrak{P} \subset \Lambda$; and the quotient $\Lambda/\mathfrak{P}$ is called the *residue class field* $\bar{F}$.

LEMMA 11.5. *The formula* $d_v(x, y) = (-1)^{v(x)v(y)} x^{v(y)}/y^{v(x)} \bmod \mathfrak{P}$ *defines a continuous Steinberg symbol* d_v *on* F *with values in the discrete group* $\bar{F}^\bullet = (\Lambda/\mathfrak{P})^\bullet$.

(Compare Serre, *Corps locaux,* p. 217.) This d_v is called the *tame symbol* associated with the valuation v. Evidently d_v gives rise to a homomorphism from K_2F onto the group $\bar{F}^\bullet = K_1(\bar{F})$.

Proof of 11.5. The element $\pm x^{v(y)}/y^{v(x)}$ is a unit of Λ, since both $x^{v(y)}$ and $y^{v(x)}$ have the same image (namely $v(x)v(y)$) under v. It is clear that d_v is bimultiplicative, and continuous in the v-topology. The proof that $d_v(1-x, x) = 1$ will be divided into several cases. If $v(x) > 0$, then $x \in \mathfrak{P}$, hence $1-x \equiv 1 \bmod \mathfrak{P}$ and $v(1-x) = 0$, so that

$$(-1)^{v(1-x)v(x)}(1-x)^{v(x)}/x^{v(1-x)} = (1-x)^{v(x)} \equiv 1 \bmod \mathfrak{P}.$$

The proof when $v(1-x) > 0$ is similar. Now suppose that $v(x) < 0$. Then $x^{-1} \in \mathfrak{P}$, hence the quotient

$$(1-x)/x = -1 + x^{-1} \equiv -1 \bmod \mathfrak{P}$$

is a unit. Therefore $v(1-x) = v(x)$, and

$$(1-x)^{v(x)}/x^{v(1-x)} = ((1-x)/x)^{v(x)} \equiv (-1)^{v(x)} \bmod \mathfrak{P}.$$

Multiplying by the sign $(-1)^{v(1-x)v(x)} = (-1)^{v(x)}$, we obtain $1 \bmod \mathfrak{P}$, as required. The case $v(1-x) < 0$ is similar. Since the remaining case $v(x) = v(1-x) = 0$ is trivial, this proves 11.5. ∎

Gauss and Quadratic Reciprocity

To illustrate these concepts let us look at the field Q of rational numbers. What Steinberg symbols $c(x,y)$ can be defined on the field Q?

For any prime p, the p-adic valuation v_p on Q gives rise to a Steinberg symbol $d_{v_p}(x,y)$ with values in the cyclic group $(Z/pZ)^\bullet$ of order $p-1$. If p is odd we will denote this symbol briefly by $(x,y)_p$, and its target group $(Z/pZ)^\bullet$ by A_p.

For $p = 2$ this construction is useless. However a 2-adic symbol $(x,y)_2$ can be defined as follows. Any non-zero rational can be written uniquely as a product of the form $\pm 2^j 5^k u$, where k equals 0 or 1, and where u is a quotient of integers congruent to 1 modulo 8. Now if

$$x = (-1)^i 2^j 5^k u, \quad y = (-1)^I 2^J 5^K u',$$

then set

$$(x,y)_2 = (-1)^{iI+jK+kJ}.$$

Thus the target group A_2 is the cyclic group $\{\pm 1\}$. The verification that this is a well defined Steinberg symbol will be left as an exercise.

REMARK. The following assertion may help to motivate the definition of $(x,y)_p$.

> *For any prime* p *suppose that a Steinberg symbol* $c : Q^\bullet \times Q^\bullet \to A$, *with values in a Hausdorff topological group* A, *is continuous with respect to the* p-*adic topology on* $Q^\bullet$. *Then there is one and only one homomorphism from* A_p *to* A *which carries the symbol* $(x,y)_p$ *to* $c(x,y)$ *for every* x *and* y.

Briefly speaking, $(x,y)_p$ is the "universal continuous Steinberg symbol" for the p-adic topology on $Q^\bullet$. This statement is a special case of a much more general theorem, due to Calvin Moore, which is proved in the Appendix.

Here is an outline of the proof. Let p^n be any prime power which is greater than 2. Then the congruence

(4) $$(1-rp^n)^p \equiv 1-rp^{n+1} \pmod{p^{n+2}}$$

follows easily from the binomial theorem. Now suppose that p is odd, and that r is prime to p. Let u_1 denote any quotient of the form s/t with $s \equiv t \equiv 1 \pmod{p}$. Using (4), we note that u_1 can be approximated arbitrarily closely, in the p-adic topology, by a power of 1-rp. In fact we can first choose i so that

$$(1-rp)^i t \equiv s \pmod{p^2},$$

then choose j so that

$$(1-rp)^{i+jp} t \equiv s \pmod{p^3},$$

and so on.

Since $c(rp, (1-rp)^i) = 1$ for every exponent i, it follows by continuity that $c(rp, u_1) = 1$ for every such $u_1 = s/t$. But the entire multiplicative group $Q^\bullet$ is generated by such products rp, with r relatively prime to the fixed prime p. Thus we have proved that

(5) $$c(x,u_1) = 1 \quad \text{for all } x \text{ in } Q^\bullet.$$

If r and r' denote integers prime to p, then it follows immediately from (5) that $c(r,r')$ depends only on the residue classes of r and r' modulo p. But, applying Steinberg's theorem that every symbol on a finite field must be trivial (§9.9), this proves that

(6) $$c(r,r') = 1.$$

Let λ denote a primitive root modulo p. Then any x and y in $Q^\bullet$ can be written more or less uniquely in the form

$$x = p^i \lambda^j u_1, \quad y = p^I \lambda^J u_1';$$

and it follows that

$$c(x,y) = c(p,p)^{iI} c(\lambda,p)^{jI-iJ}.$$

Since the equalities

$$c(\lambda,p)^{p-1} = c(\lambda^{p-1},p) = 1$$

and

$$c(p,p) = c(-1,p) = c(\lambda,p)^{(p-1)/2}$$

follow from (5), the proof for p odd can now easily be completed.

For $p = 2$ a similar argument shows that every number u which can be expressed as a quotient s/t with $s \equiv t \equiv 1 \pmod 8$ can be approximated arbitrarily closely, in the 2-adic topology, by a power of 9. Using the equalities

$$\begin{aligned} c(9,-1) &= c(3,-1)^2 = c(3,(-1)^2) = 1, \\ c(9,-2) &= c(3,-2)^2 = 1, \text{ and} \\ c(9,3) &= c(-3,3)^2 = 1, \end{aligned}$$

it follows by continuity that

$$c(u,-1) = c(u,-2) = c(u,3) = 1$$

for every such u. Since -1, -2, and 3 generate a subgroup of $Q^{\bullet}$ which is everywhere dense, this proves that

$$c(u,x) = 1 \quad \text{for all } x. \tag{7}$$

As an example, taking $u = -5/3$, it follows that

$$c(5,x) = c(-3,x).$$

Taking $x = 4$, we see that $c(5,4) = 1$, hence $c(5,-1) = c(5,-4) = 1$, and therefore

$$c(5,5) = c(5,-1) = 1. \tag{8}$$

Similarly the equation $c(-5,-1) = c(3,x)$ for $x = -2$ implies that $c(-5,-2) = 1$, and hence

$$c(5,2) = c(-1,-1). \tag{9}$$

Now combining (7), (8), and (9) with the evident equation $c(2,2) = c(2,-1) = 1$, we see that

$$c((-1)^i 2^j 5^k u, (-1)^I 2^J 5^K u') = c(-1,-1)^{iI + jK + kJ};$$

which clearly completes the proof. ∎

Using these Steinberg symbols $(x,y)_p$, we are now ready to compute the group K_2Q.

THEOREM 11.6 (Tate). *The group* K_2Q *is canonically isomorphic to the direct sum* $A_2 \oplus A_3 \oplus A_5 \oplus \ldots$, *where* A_2 *is the cyclic group* $\{\pm 1\}$, *and where* $A_p = (Z/pZ)^{\bullet}$ *for* p *odd.*

In fact the isomorphism will be given by the correspondence

$$\{x,y\} \mapsto (x,y)_2 \oplus (x,y)_3 \oplus (x,y)_5 \oplus \ldots$$

for all x and y in $Q^\bullet$.

Tate remarks that his proof of this theorem is lifted directly from the argument which was used by Gauss in his first proof of the quadratic reciprocity law. (Compare Gauss, *Disquisitiones Arithmeticae,* Yale Univ. Press 1966, pp. 84-98.)

To start the proof, for each positive integer m let L_m denote the subgroup of K_2Q generated by all symbols $\{x,y\}$ where x and y are integers of absolute value $\leq m$. Then clearly

$$L_1 \subset L_2 \subset L_3 \subset \ldots$$

with union K_2Q. Note that $L_m = L_{m-1}$ if m is not a prime number.

LEMMA 11.7. *For each prime* p *the quotient group* L_p/L_{p-1} *is cyclic of order* $p-1$.

In particular the quotient L_2/L_1 is trivial. Assuming this lemma for the moment, the proof proceeds easily as follows.

For each prime p the correspondence $\{x,y\} \mapsto (x,y)_p$ defines a homomorphism from K_2Q to A_p. If p is odd, it is clear that this homomorphism annihilates L_{p-1}, but maps L_p onto the cyclic group $A_p = (Z/pZ)^\bullet$. Hence it induces an isomorphism $L_p/L_{p-1} \cong A_p$. On the other hand, for $p = 2$, this homomorphism maps the generator $\{-1,-1\}$ of L_1 onto the element $(-1,-1)_2 = -1$, and hence induces an isomorphism from $L_1 = L_2$ to A_2. An easy induction now shows that, for each prime p, the correspondence

$$\{x,y\} \mapsto (x,y)_2 \oplus (x,y)_3 \oplus \ldots \oplus (x,y)_p$$

maps the group L_p isomorphically onto the direct sum $A_2 \oplus A_3 \oplus \ldots \oplus A_p$. Taking the direct limit as $p \to \infty$, the Theorem follows.

To prove Lemma 11.7, consider the correspondence

$$\phi : (Z/pZ) \to L_p/L_{p-1}$$

defined by the formula

$$x \mapsto \{x,p\} \text{ modulo } L_{p-1}\,.$$

Here x is to vary over all non-zero integers of absolute value less than p. To show that ϕ is well defined, and a homomorphism, we suppose that

$$xy \equiv z \mod p,$$

where x, y and z are all non-zero integers of absolute value less than p. Then $xy = z+pr$ with $|pr| \le |xy| + |z| \le (p-1)^2 + p-1$, hence $|r| < p$. Now

$$1 = z/xy + pr/xy$$

so

$$1 = \{z/xy, pr/xy\} \equiv \{z/xy, p\} \mod L_{p-1}.$$

Therefore

$$\{z,p\} \equiv \{xy,p\} \mod L_{p-1},$$

so that ϕ is a homomorphism, and (taking $y = 1$) ϕ is well defined.

To prove that ϕ is surjective, note that L_p is generated by the symbols $\{x,\pm p\}$, $\{\pm p,x\}$, and $\{\pm p,\pm p\}$, together with L_{p-1}. Hence the identities

$$\{-p,-p\} \equiv \{p,p\} \equiv \phi(-1) \mod L_{p-1},$$

$$\{\pm p,x\}^{-1} = \{x,\pm p\} \equiv \phi(x) \mod L_{p-1},$$

and

$$\{-p,p\} = \{p,-p\} = 1,$$

show that ϕ is indeed surjective. This proves that L_p/L_{p-1} has at most $p-1$ elements. Since we already know, using the symbol $(x,y)_p$, that L_p/L_{p-1} has at least $p-1$ elements, this completes the proof. ∎

Another way of stating our conclusion is the following.

COROLLARY 11.8. *Given any Steinberg symbol* c(x,y) *on the rational numbers, with values in an abelian group* A, *there exist unique homomorphisms*

$$\phi_p : A_p \to A$$

so that

$$c(x,y) = \prod \phi_p((x,y)_p),$$

the product being taken over all prime numbers p.

In this formulation, the result could have been proved directly, without ever mentioning K_2.

To illustrate this corollary, consider the local symbol $(x,y)_\infty$, defined by

$$(x,y)_\infty = \begin{cases} +1 & \text{if } x>0 \text{ or } y>0 \\ -1 & \text{if } x,y<0, \end{cases}$$

which is associated with the embedding of the rational numbers in the real numbers. (Compare §8.4.) This is the "universal continuous Steinberg symbol" for the archimedean topology of Q. According to 11.8 there must be a relation of the form

$$(x,y)_\infty = \prod \phi_p((x,y)_p).$$

In fact one has the following.

QUADRATIC RECIPROCITY LAW. *The symbol* $(x,y)_\infty$ *is equal to the product, over all primes* p, *of* $((x,y))_p$, *where the Hilbert symbol* $((x,y))_p = \pm 1$ *is defined to be* $(x,y)_2$ *if* $p = 2$ *and is defined by the condition*

$$((x,y))_p \equiv (x,y)_p{}^{(p-1)/2} \mod p$$

if p *is odd.*

Proof. It is clear from the Corollary that there exists some relation of the form

$$(x,y)_\infty = \prod ((x,y))_p{}^{\varepsilon_p},$$

where the exponents $\varepsilon_2, \varepsilon_3, \varepsilon_5, \ldots$ must be either 0 or 1. Taking $x = y = -1$ we see that the exponent ε_2 must be 1. If p is a prime of the form $8k\pm 3$, then since

$$(2,p)_\infty = 1, \quad (2,p)_2 = -1,$$

we must have

$$(2,p))_p{}^{\varepsilon_p} = -1,$$

so that ε_p cannot be zero. Similarly, if p is a prime of the form $8k+7$ (or $8k+3$), then the equations

$$(-1,p)_\infty = 1 \qquad (-1,p)_2 = -1$$

imply that ε_p cannot be zero.

There remains only the case of a prime of the form $8k+1$. Following Gauss we prove the following.

LEMMA 11.9. *If* p *is a prime of the form* 8k+1, *then there exists a prime* $q < \sqrt{p}$ *so that* p *is not a quadratic residue modulo* q.

(Examples such as $109 \equiv 2^2 \bmod 3\cdot5\cdot7$ show that the hypothesis $p \equiv 1 \bmod 8$ is essential, at least for small values of p.)

Proof (following Tate). Consider the product

$$N = \frac{p-1^2}{4} \cdot \frac{p-3^2}{4} \cdot \frac{p-5^2}{4} \cdot \ldots \cdot \frac{p-m^2}{4}.$$

Here m should be the largest odd number less than $\sqrt{p}$, so that $m^2 < p < (m+2)^2$. Then for each factor $(p-i^2)/4$ of the product N we have

$$0 < \frac{p-i^2}{4} < \frac{(m+2)^2-i^2}{4} = \frac{m+2+i}{2} \cdot \frac{m+2-i}{2} .$$

Taking the product, for $i = 1,3,5,\ldots,m$, this yields

$$0 < N < (m+1)!\ .$$

Now suppose that p is a quadratic residue modulo every prime less than $\sqrt{p}$. Then we will prove that

$$N \equiv 0 \mod (m+1)!\ ,$$

thus yielding a contradiction. We will use the notation $[\xi]$ for the largest integer $\leq \xi$.

First note, following Gauss, that in order to prove a congruence of the form $a_1a_2\ldots a_k \equiv 0 \bmod n!$ it suffices to prove, for each prime power $q^s \leq n$, that at least $[n/q^s]$ of the factors a_j are divisible by q^s. The congruence then follows easily, using the identity $n! = \prod_{q^s \leq n} q^{[n/q^s]}$.

Thus in our case, for each prime power $q^s \leq m+1$, we must prove that at least $[(m+1)/q^s]$ of the numbers $(p-i^2)/4$ are divisible by q^s. In other words we must show that the congruence

$$p \equiv i^2 \mod 4q^s$$

has at least $[(m+1)/q_s]$ solutions in the interval $0 < i < m+1$.

First we will show that p is indeed a quadratic residue modulo $4q^s$. Since $p \equiv 1 \bmod 8$, it is known that p is a quadratic residue modulo any power of 2. So it suffices to consider the case q odd, hence $q^s \neq m+1$.

Then

$$q \leq q^s \leq m < \sqrt{p},$$

so p is a residue modulo q; and it follows easily that p is a residue modulo $4q^s$.

Thus the congruence $p \equiv i^2 \mod 4q^s$ has at least one solution i. Now, changing the sign of i if necessary, and adding a multiple of $2q^s$, we obtain a solution i_0 which lies in the interval $0 < i_0 < q^s$. (This is possible since $(i+2q^s)^2 \equiv i^2 \mod 4q^s$.) Similarly we obtain a solution $2q^s - i_0$ lying between q^s and $2q^s$, a solution $i_0 + 2q^s$ between $2q^s$ and $3q^s$, and so on. Thus, for each positive n, there exist at least $[n/q^s]$ solutions between 0 and n. Taking $n = m+1$, this completes the proof of Lemma 11.9. ∎

The proof of the quadratic reciprocity law, following Gauss and Tate, can now be completed as follows. Suppose that p is a non-residue modulo q, where $q < p$ and $p \equiv 1 \mod 8$. We may suppose inductively that the exponent ε_q equals 1. Then $(p,q)_\infty = ((p,q))_2 = 1$ but $((p,q))_q = -1$. So it follows that $((p,q))_p^{\varepsilon_p} = -1$, and hence that $\varepsilon_p \neq 0$. This completes the proof. ∎

Remark. Let F(x) denote the field of rational functions

$$f = (a_0x^n+\ldots+a_n)/(b_0x^m+\ldots+b_m)$$

in one variable over F. It will be convenient to set

$$\deg f = n-m, \quad \text{lead coef } f = a_0/b_0.$$

The technique used above to compute K_2Q can also be applied to $K_2F(x)$, and yields a split exact sequence

$$1 \to K_2F \to K_2F(x) \to \bigoplus (F[x]/\mathfrak{p})^\bullet \to 1,$$

where $\mathfrak{p}$ ranges over all non-zero prime ideals in the polynomial ring. (To prove that the sequence splits one uses a symbol such as $c(f,g) = \{\text{lead coef } f, \text{lead coef } g\}$ with values in K_2F.)

Just as in the rational number case, the proof is based on the symbols $(f,g)_\mathfrak{p}$ associated with the various $\mathfrak{p}$-adic valuations on F(x). And just as in the rational case, one valuation is conspicuously absent from the list. In this case it is the valuation

$$v_\infty(f) = -\deg(f)$$

associated with the point at infinity. Hence, just as before, we can derive a formula which expresses the corresponding Steinberg symbol

$$(f,g)_\infty = (-1)^{\deg f \deg g}(\text{lead coef } g)^{\deg f}/(\text{lead coef } f)^{\deg g}$$

in terms of the $(f,g)_{\mathfrak{p}}$. The appropriate formula, due to Weil, is

$$(f,g)_\infty^{-1} = \prod \text{norm } (f,g)_{\mathfrak{p}} ,$$

taking the product over all non-zero prime ideals $\mathfrak{p}$, and using the norm homomorphism from $(F[x]/\mathfrak{p})^\bullet$ to $F^\bullet$. (Compare Bass, *Algebraic K-Theory*, p. 333.) If f and g are relatively prime polynomials, then the right side of this equation can be written as

$$\prod_{g(\xi)=0} f(\xi) / \prod_{f(\eta)=0} g(\eta) ,$$

where ξ and η range over the algebraic closure of F, and n-fold zeros are to be counted n times.

Uncountable Fields

To conclude this section we will give one more application of Lemma 11.5.

THEOREM 11.10. *If a field* F *has uncountably many elements, then* K_2F *is uncountable also.*

Proof. Let $\Pi \subset F$ be the prime field, and let $X = \{x_\alpha\}$ be a maximal set of algebraically independent elements over Π. Thus F is an algebraic extension of the uncountable function field $\Pi(X)$. Choosing one of the indeterminates $x_0 \in X$ and letting $X' = X - \{x_0\}$, we obtain a discrete valuation on $\Pi(X)$, with residue class field $\Pi(X')$, by considering the place $f(x_0) \mapsto f(0)$. (Here we are thinking of $f(x_0)$ as a polynomial in the indeterminate x_0 with coefficients in $\Pi(X')$.) Extend this place to a place on F with values in the algebraic closure of $\Pi(X')$. (Compare

Lang, *Introduction to Algebraic Geometry*, p. 8.) Then for every finite extension E of $\Pi(X)$ within F we obtain a discrete valuation on E whose residue class field $\bar{E}$ is a finite extension of $\Pi(X')$. Map K_2E to $\bar{E}^\bullet$ by 11.5. If E_1 is an extension field of E with ramification index r, then it is easily verified that the following diagram is commutative,

$$\begin{array}{ccc} K_2E & \longrightarrow & K_2E_1 \\ \downarrow & & \downarrow \\ \bar{E}^\bullet & \xrightarrow{\ r\ } & \bar{E}_1^\bullet \end{array}$$

where r denotes the homomorphism $e \mapsto e^r$. In order to make this bottom homomorphism injective, we will divide out by the countable subgroup consisting of all roots of unity in $\bar{E}^\bullet$. Thus we obtain

$$\begin{array}{ccc} K_2E & \longrightarrow & K_2E_1 \\ \downarrow & & \downarrow \\ \bar{E}^\bullet/(\text{roots of unity}) & \to & \bar{E}_1^\bullet/(\text{roots of unity}), \end{array}$$

where the bottom arrow is now an injection.

Passing to the direct limit as E varies over all finite extensions of $\Pi(X)$ in F, we thus obtain a homomorphism from K_2F onto a direct limit group which contains $\bar{E}^\bullet/(\text{roots of unity})$ for all such E. This proves that the group K_2F is necessarily uncountable. ∎

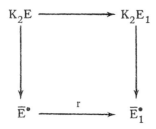

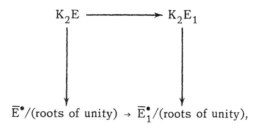

§12. Proof of Matsumoto's Theorem

Let c be a Steinberg symbol on the field F with values in a multiplicative abelian group A. (Compare §11.3.) We will use c to construct a central extension

$$1 \to A \to G \to SL(n,F) \to 1.$$

Here n could be any positive integer, but for convenience we assume that $n \geq 3$. The extension will be constructed first over the subgroup D of diagonal matrices, then over the larger group M of monomial matrices, and finally over the entire group $SL(n,F)$.

To construct the preliminary extension

$$1 \to A \to H \to D \to 1,$$

let H be the set $D \times A$ with the following product operation. If $d = \mathrm{diag}(u_1,\ldots,u_n)$ and $d' = \mathrm{diag}(v_1,\ldots,v_n)$ then

$$(d,a)\,(d',a') = (dd', aa' \prod_{i \geq j} c(u_i,v_j)).$$

It is easily verified that this product is associative, and hence makes H into a group. Let

$$\phi : H \to D$$

be the projection to the first factor. Thus ϕ is a homomorphism with kernel $I \times A$ contained in the center of H. We will identify this kernel with A. Commutators in H can be computed just as in §8.3:

LEMMA 12.1. *If* $\phi(h) = \mathrm{diag}(u_1,\ldots,u_n)$ *and* $\phi(k) = \mathrm{diag}(v_1,\ldots,v_n)$, *then* $hkh^{-1}k^{-1}$ *is equal to the product*

$$c(u_1,v_1)\, c(u_2,v_2) \,\ldots\, c(u_n,v_n).$$

Proof. This follows easily, using the skew-symmetry of c and the equation $u_1 \ldots u_n = v_1 \ldots v_n = 1$. ∎

Now given integers $i \neq j$ between 1 and n let us define symbols $h_{ij}(u) \in H$. If $i < j$ we define $h_{ij}(u)$ to be the pair $(d_{ij}(u),1)$ in $D \times A = H$ where $d_{ij}(u)$ denotes the matrix

$$\mathrm{diag}(1,\ldots,1,u,1,\ldots,1,u^{-1},1,\ldots,1)$$

with entry u in the i-th place and entry u^{-1} in the j-th place. If $i > j$ we define $h_{ij}(u)$ to be the pair $(d_{ij}(u), c(u,u))$.

LEMMA 12.2. *These symbols* $h_{ij}(u)$ *satisfy the identities* $h_{ji}(u) = h_{ij}(u)^{-1} = h_{ik}(u)^{-1}h_{kj}(u)^{-1}$ *and*

$$h_{ij}(u)\,h_{ij}(v) = c(u,v)\,h_{ij}(uv),$$

just as in §9.

The proof is straightforward.

Next we want to construct a group W containing H and mapping onto the group M of monomial matrices. As a first step we will construct a subgroup W_0, generated by certain symbols $w_{ij}(\pm 1)$. There are two possibilities.

Case 1. If $c(-1,-1) = 1$ we define W_0 to be the group M_0, consisting of monomial matrices in SL(n,F) whose non-zero entries are all ± 1. We define $w_{ij}(1) = w_{ji}(-1)$ to be that monomial matrix with ij-th entry 1 and ji-th entry -1, whose other non-zero entries are 1's along the diagonal. We define $\phi_0 : W_0 \to M_0$ to be the identity map.

Case 2. If $c(-1,-1) \neq 1$, then F must be a field of characteristic zero by 9.9. Hence F contains the ring Z of integers, and we can identify M_0 with the group of all monomial matrices in SL(n,Z). Let W_0 be the group lying over M_0 in the two-fold central extension $St(n,Z) \to SL(n,Z)$. Thus a natural map $\phi_0 : W_0 \to M_0$ is defined, and elements $w_{ij}(\pm 1) \in W_0$ are defined as in §10.

REMARK. Alternatively one could identify M_0 with a subgroup of the rotation group SO(n) and define W_0 to be the corresponding subgroup of its 2-fold covering Spin(n).

Next we must relate this group W_0 to the group H by defining an action of W_0 on H. Since the action of an element w_0 will depend only on $\phi_0(w_0)$, we will work first with monomial matrices.

For each permutation π of $\{1,\ldots,n\}$ let p_π denote the associated permutation matrix. Every monomial matrix m can be written uniquely as the product $p_\pi \operatorname{diag}(u_1,\ldots,u_n)$ of a permutation matrix and a diagonal matrix, within $GL(n,F)$.

LEMMA 12.3. *For each monomial matrix* $m = p_\pi \operatorname{diag}(u_1,\ldots,u_n)$ *there is one and only one automorphism* $\alpha(m)$ *of* H *which leaves* A *pointwise fixed and carries each* $h_{ij}(v)$ *to* $c(u_i u_j^{-1},v)\, h_{\pi(ij)}(v)$. *This automorphism* $\alpha(m)$ *depends homomorphically on* m *and coincides with the inner automorphism* $h \mapsto h_1 h h_1^{-1}$ *whenever* $m = \phi(h_1)$ *is a diagonal matrix.*

Proof. If $u_1 \ldots u_n = +1$ then we see by 12.1 that the correspondence

$$h_{ij}(v) \mapsto c(u_i u_j^{-1},v)\, h_{ij}(v)$$

is an inner automorphism of H, hence certainly an automorphism. If $u_1 \ldots u_n = -1$ then this correspondence extends to an inner automorphism of the larger group $H(n+1)$ associated with $SL(n+1,F)$. So it defines an automorphism of H in this case also. Let us denote this automorphism temporarily by $\beta(u_1,\ldots,u_n)$.

For any permutation π of $\{1,\ldots,n\}$ the correspondence $h_{ij}(v) \mapsto h_{\pi(ij)}(v)$ preserves all of the relations 12.2 between the $h_{ij}(v)$ and preserves the commutator relations 12.1. Therefore it also gives rise to an automorphism of H leaving A pointwise fixed. Let us call this automorphism β_π; and let $\alpha(m) = \beta_\pi\, \beta(u_1,\ldots,u_n)$.

Now noting the relation

$$\beta(u_1,\ldots,u_n)\,\beta_\pi = \beta_\pi \beta(u_{\pi(1)},\ldots,u_{\pi(n)}),$$

the proof is easily completed. ∎

In the special case $m = \phi_0(w_0)$ we will use the notation $h \mapsto w_0 h w_0^{-1}$ for this automorphism $\alpha(m)$.

In order to construct a group W, generated by H and W_0, we also need to describe the intersection $H \cap W_0$. Let $H_0 \subset H$ be the subgroup generated by the symbols $h_{ij}(-1)$. This can be identified with the subgroup of W_0 generated by corresponding symbols $d_{ij}(-1)$, in the case $W_0 \cong M_0$; or by corresponding symbols $h_{ij}(-1)$ in the case $W_0 \subset St(n,Z)$.

Definition of W. Let W be the identification space of $H \times W_0$ which is obtained by identifying (hh_0, w_0) with $(h, h_0 w_0)$ for every $h_0 \in H_0$. Define the product operation in W by

$$(h,w_0)(h',w_0') = (h(w_0 h' w_0^{-1}), w_0 w_0'),$$

using the action of W_0 on H described in 12.3.

LEMMA 12.4. *This product operation is compatible with the identifications, and makes the set* W *into a group. Furthermore, defining* $\phi : W \to M$ *by* $\phi(h,w_0) = \phi(h)\phi_0(w_0)$, *the group* W *is a central extension of* M *with kernel isomorphic to* A.

The proof is straightforward.

In practice we will identify the groups $A \subset H$ and W_0 with their isomorphic images in W. Note then that $H \cap W_0 = H_0$.

In order to work effectively with this group W we will need an explicit description of inner automorphisms of W.

LEMMA 12.5. *Suppose that* $\phi(w) = p_\pi \operatorname{diag}(u_1,\ldots,u_n)$.

Then

$$w h_{ij}(v) w^{-1} = c(u_i u_j^{-1}, v)\, h_{\pi(ij)}(v)$$

and

$$w w_{ij}(1) w^{-1} = h_{\pi(ij)}(u_i u_j^{-1})\, w_{\pi(ij)}(1).$$

Proof. We already know how to compute $wh_{ij}(v)w^{-1}$ in the special case where w belongs to either W_0 or H. Since the group W is generated by W_0 and H, it is not difficult to verify the above formula for $wh_{ij}(v)w^{-1}$ for arbitrary w.

Next let us conjugate $w_{ij}(1)$ by an element h, with $\phi(h) = \operatorname{diag}(u_1,\ldots,u_n)$. Express h as a product

$$h = h_{ir}(u_i)h_{jr}(u_j)h'$$

for some r, where h' commutes with $w_{ij}(1)$, and note that

$$w_{ij}(1)hw_{ij}(-1) = h_{jr}(u_i)c(-1,u_i)h_{ir}(u_j)h'.$$

Hence the commutator $[h,w_{ij}(1)]$ is equal to

$$h_{ir}(u_i)h_{jr}(u_j)h_{ir}(u_j)^{-1}c(-1,u_i)h_{jr}(u_i)^{-1}.$$

Using 12.1 and 12.2 and a little patience, this expression can now be transformed into $h_{ij}(u_iu_j^{-1})$. Hence $hw_{ij}(1)h^{-1} = h_{ij}(u_iu_j^{-1})w_{ij}(1)$, as asserted. Since we can conjugate $w_{ij}(1)$ by an element of W_0 using 9.4, the proof is now easily completed. ∎

In order to pass from the subgroup M of monomial matrices to the full group $SL(n,F)$ we will use a variant of the Bruhat normal form for a matrix. (Compare §9.15.) Let T denote the group of upper triangular matrices with zeros below the diagonal and ones on the diagonal.

LEMMA 12.6. *Every matrix* s *in* $SL(n,F)$ *can be written as a product* tmt' *with* t *and* t' *in* T *and* m *in* M. *Although* t *and* t' *may not be uniquely determined, the monomial matrix* m *is uniquely determined by* s. *Thus a well defined retraction* $\rho : SL(n,F) \to M$ *is defined by the formula* $\rho(tmt') = m$.

(Of course ρ is not a homomorphism.)

Proof of 12.6. The first statement follows immediately from 9.15, or can be proved by using suitable elementary row and column operations. To prove the second, suppose that

$$t_1mt_2 = t_3m't_4.$$

Multiplying on the left by t_3^{-1} and on the right by t_2^{-1}, it follows that

$$tm = m't'$$

for suitable t,t'.

Now examine the standard formula which expresses the determinant of the matrix tm as a sum of $n!$ monomials. Inspection shows that precisely one of these monomials is non-zero: namely the monomial whose n factors are precisely the n non-zero entries of the matrix m. Since a similar statement holds for the product $m't'$, this proves that $m = m'$. ∎

To simplify the notation, it will be convenient to use a single letter α to denote a pair of consecutive indices $(i,i+1)$. The reversed pair $(i+1,i)$ will be denoted by $-\alpha$. Thus $d_\alpha(u)$ stands for the diagonal matrix with entry u in the i-th place, u^{-1} in the $(i+1)$-st place and 1's along the diagonal otherwise. Similarly $m_\alpha(u)$ stands for the monomial matrix with entry u in the α-th place, $-u^{-1}$ in the $-\alpha$-th place, and 1's along the diagonal otherwise.

Although the function $\rho : SL(n,F) \to M$ is not a homomorphism, note that it does satisfy the condition

$$\rho(ds) = d\rho(s), \quad \rho(sd) = \rho(s)d$$

for any diagonal matrix d in $SL(n,F)$. The following properties of ρ will be particularly important.

LEMMA 12.7. *For each* s *in* $SL(n,F)$ *the expression* $\rho(m_\alpha(1)s)$ *is equal either to* $m_\alpha(1)\rho(s)$, *or to* $d_\alpha(u)^{-1}\rho(s)$ *for some uniquely determined element* u *of* $F^\bullet$. *Similarly* $\rho(sm_\beta(-1))$ *is equal either to* $\rho(s)m_\beta(-1)$ *or to* $\rho(s)d_\beta(v)$ *for some* v.

REMARK. We are being careful to write these formulas in such a manner that they remain true in the Steinberg group, with w_α in place of m_α and h_α in place of d_α.

Proof of 12.7. (Compare §9.15.) In the expression $s = tmt'$ we can write t' uniquely as a product $e_\beta^v t_\beta$, where e_β^v denotes the elementary matrix with β-th entry equal to v, and where t_β stands for any matrix in T whose β-th entry is zero. Then

$$sm_\beta(-1) = tme_\beta^v m_\beta(-1)\, t'_\beta \tag{1}$$

for some t'_β. If v happens to be zero, this shows already that $\rho(sm_\beta(-1)) = \rho(s)\, m_\beta(-1)$.

Suppose then that $v \neq 0$. Let π be the permutation which is associated with the monomial matrix m, and let $\beta = (j,j+1)$. If $\pi(j) < \pi(j+1)$, so that the conjugate $me^{v}_{\beta}m^{-1}$ belongs to T also, then we can push the factor e^{v}_{β} to the left in equation (1) and conclude again that $\rho(sm_{\beta}(-1)) = \rho(s)\,m_{\beta}(-1)$.

If $\pi(j) > \pi(j+1)$, then substitute the identity

$$e^{v}_{\beta} = e^{v^{-1}}_{-\beta}\, e^{-v}_{\beta} m_{\beta}(v)$$

into (1), and push the factor $e^{v^{-1}}_{-\beta}$ to the left and the factor e^{-v}_{β} to the right. It follows that

$$sm_{\beta}(-1) = t''mm_{\beta}(v)\,m_{\beta}(-1)\,t'''.$$

Since $m_{\beta}(v)\,m_{\beta}(-1) = d_{\beta}(v)$, this proves that $\rho(sm_{\beta}(-1)) = \rho(s)d_{\beta}(v)$ in this case.

The computation of $\rho(m_{\alpha}(1)\,s)$ is similar. Let $s = t_{\alpha}e^{-u}_{\alpha}mt'$, so that

$$m_{\alpha}(1)\,s = t'_{\alpha}m_{\alpha}(1)\,e^{-u}_{\alpha}mt'.$$

Then if $\pi^{-1}(i) < \pi^{-1}(i+1)$, we can shift the factor e^{-u}_{α} to the right, and conclude that $\rho(m_{\alpha}(1)\,s) = m_{\alpha}(1)\,\rho(s)$. On the other hand if $u \neq 0$ and $\pi^{-1}(i) > \pi^{-1}(i+1)$, then substituting $e^{-u}_{\alpha} = m_{\alpha}(-u)\,e^{u}_{\alpha}e^{-u^{-1}}_{-\alpha}$, and pushing the two elementary matrices to the left and right, we obtain

$$m_{\alpha}(1)\,s = t''m_{\alpha}(1)\,m_{\alpha}(-u)\,mt'''.$$

Since $m_{\alpha}(1)\,m_{\alpha}(-u) = d_{\alpha}(u)^{-1}$, this completes the proof of 12.7. ∎

REMARK. For later use, we will need the equations which were used above in a slightly more explicit form. To each monomial matrix m and each index pair β associate the field element $f(\beta,m)$ which is defined by the equation

$$me^{x}_{\beta}m^{-1} = e^{f(\beta,m)\,x}_{\pi(\beta)}.$$

Then if $\phi(w) = m$, it follows that $wh_{\beta}(v)\,w^{-1} = c(f(\beta,m),\ v)\,h_{\pi(\beta)}(v)$. (Compare §§9.4, 12.5.) Note that $f(-\beta,m) = f(\beta,m)^{-1}$.

Now if $s = tme^{v}_{\beta}t_{\beta}$, with $v \neq 0$, then either

(2) $$sm_{\beta}(-1) = te^{f(\beta,m)\,v}_{\pi(\beta)}mm_{\beta}(-1)\,t''$$

or

(3) $$sm_\beta(-1) = te_{\pi(-\beta)}^{f(-\beta,m)v^{-1}} md_\beta(v)t''$$

is the required equation. Similarly, if $s = t_\alpha e_\alpha^{-u} mt$ with $u \neq 0$, and if $\alpha = \pi(\gamma)$, then either

(4) $$m_\alpha(1)s = t' m_\alpha(1) me_\gamma^{-uf(\gamma,m)^{-1}} t$$

or

(5) $$m_\alpha(1)s = t' d_\alpha(u)^{-1} me_{-\gamma}^{-f(\gamma,m)u^{-1}} t$$

is the required equation.

We are now ready to give Matsumoto's ingeneous construction of a central extension of $SL(n,F)$. Let

$$X \subset SL(n,F) \times W$$

be the set of all pairs (s,w) satisfying the condition

$$\rho(s) = \phi(w).$$

Since there is no obvious way of making this set X into a group, we do something else instead. Let G be the group of permutations of X which is generated by certain permutations $\lambda(h)$, $\mu(t)$, and η_α, defined as follows.

For each $h \in H$ let $\lambda(h)$ denote the permutation

$$\lambda(h)(s,w) = (\phi(h)s, hw)$$

of the set X. It is clear that this transformation preserves the condition $\rho(s) = \phi(w)$, and that λ maps the group H injectively into the group of all permutations of X.

For each $t \in T$ let $\mu(t)$ denote the permutation

$$\mu(t)(s,w) = (ts,w).$$

Clearly μ maps T injectively into the group of all permutations of X.

For each $\alpha = (i,i+1)$ let η_α denote the permutation defined by setting $\eta_\alpha(s,w)$ equal to either $(m_\alpha(1)s, w_\alpha(1)w)$ or $(m_\alpha(1)s, h_\alpha(u)^{-1}w)$ according as $\rho(m_\alpha(1)s)$ is equal to $m_\alpha(1)\rho(s)$ or $d_\alpha(u)^{-1}\rho(s)$. Clearly this definition is concocted so as to preserve the condition $\rho(s) = \phi(w)$. It is not difficult to check that

$$\eta_\alpha \eta_\alpha = \lambda(h_{-\alpha}(-1)),$$

which shows that η_α is indeed a permutation. (The identity $c(u,-u) = 1$ is used here.)

These permutations $\lambda(h)$, $\mu(t)$, and η_α must certainly generate some subgroup G of the group of all permutations of X. The key lemma is now the following.

LEMMA 12.8. *This group* G *operates in a simply transitive manner on* X.

In other words, given any (s,w) and (s′,w′) in X, there is one and only one $g \in G$ with $g(s,w) = (s',w')$.

The proof of transitivity is comparatively easy. As in the proof of 9.15, we note that SL(n,F) is generated by T and the elements $m_\alpha(1)$. So operating on (s,w) by some sequence of the permutations $\mu(t)$ and η_α we can certainly transform the first component s of (s,w) to s′. That is we can find a $g_0 \in G$ with $g_0(s,w) = (s',w^*)$. Now since both (s′,w′) and (s',w^*) belong to X, we conclude that $w' \equiv w^*$ modulo the subgroup A of W. Hence operating on (s',w^*) by a suitable $\lambda(a)$ we obtain (s′,w′).

This proves the existence of g with $g(s,w) = (s',w')$. The proof of uniqueness is more difficult, and will be given later.

Assuming the existence and uniqueness of such an element g for the moment, we easily prove the following.

THEOREM 12.9. *The group* G *is a central extension of* SL(n,F) *with kernel* $\lambda(A) \cong A$.

Proof. First note that the action of any group element g on the first component of any pair $(s,w) \in X$ is just left multiplication by some element $\Psi(g)$ of SL(n,F). This fact is true for the generators of G, and hence is true for arbitrary elements of G. This defines a homomorphism $\Psi: G \to SL(n,F)$. Since G acts transitively on X, it follows that Ψ is surjective.

The kernel of Ψ can be computed as follows. If $\Psi(g) = 1$, then g(s,w) must be a pair (s,w′). The equation $\rho(s) = \phi(w) = \phi(w')$ implies that $w' = aw$ for some $a \in A$. Thus

$$g(s,w) = \lambda(a)\,(s,w).$$

Using the simple transitivity of the action of G, this proves that $g = \lambda(a)$. Therefore the sequence

$$1 \to A \overset{\lambda}{\to} G \overset{\Psi}{\to} SL(n,F) \to 1$$

is exact.

Since inspection shows that each $\lambda(a)$ commutes with each generator of G, this proves Theorem 12.9, modulo the Lemma.

But to complete the proof we must still verify Lemma 12.8. The verification will be based on the following construction. Let G^* be the group of permutations, acting on the right of the set X, which is generated by certain permutations $\lambda^*(h)$, $\mu^*(t)$, and η^*_β constructed as follows. For each h in H let

$$(s,w)\lambda^*(h) = (s\phi(h), wh).$$

For each t in T let

$$(s,w)\mu^*(t) = (st,w).$$

Finally, for each $\beta = (j,j+1)$, define $(s,w)\eta^*_\beta$ to be either $(sm_\beta(-1), ww_\beta(-1))$ or $(sm_\beta(-1), wh_\beta(v))$ according as $\rho(sm_\beta(-1))$ is equal to $\rho(s)m_\beta(-1)$ or $\rho(s)d_\beta(v)$. (Compare 12.7.) Clearly these permutations $\lambda^*(h)$, $\mu^*(t)$, and η^*_β generate a transitive group G^* of permutations which operates on the right of the set X.

LEMMA 12.10. *Each element of the permutation group* G *commutes with each element of the permutation group* G^*.

In other words the "associative law"

$$(gx)g^* = g(xg^*)$$

is valid for every $g \in G$, $x \in X$, and $g^* \in G^*$. To prove this law, it clearly suffices to consider the special case where g is one of the generators of G and g^* is one of the generators of G^*.

If g is a generator of type $\lambda(h)$ or $\mu(t)$, or if g^* is a generator of type $\lambda^*(h)$ or $\mu^*(t)$, then this associative law can be verified without difficulty. So we will concentrate on the case $g = \eta_\alpha$ and $g^* = \eta^*_\beta$.

Note also that the first component of $\eta_\alpha(s,w)\eta^*_\beta$, with either placement of parentheses, is clearly equal to $m_\alpha(1)sm_\beta(-1)$. So we need only concentrate on the second component of $\eta_\alpha(s,w)\eta^*_\beta$.

Let $x = (s,w)$, let $s = t_\alpha e_\alpha^{-u} m e_\beta^{v} t'_\beta$, and let π be the permutation which is associated with the monomial matrix $m = \phi(w)$.

Case 1. If $\pi(\beta) \neq \pm\alpha$, then the second component of $\eta_\alpha(s,w)\eta^*_\beta$ is equal either to

$$w_\alpha(1)ww_\beta(-1) \quad \text{or} \quad h_\alpha(u)^{-1}ww_\beta(-1),$$

if $v = 0$ or $\pi(j) < \pi(j+1)$; and is equal to

$$w_\alpha(1)wh_\beta(v) \quad \text{or} \quad h_\alpha(u)^{-1}wh_\beta(v)$$

if $v \neq 0$ and $\pi(j) > \pi(j+1)$. In each of these cases, the result does not depend on the placement of parentheses!

Case 2. If $\pi(\beta) = \alpha$, then it follows from equation (2) that the second component of $\eta_\alpha(x\eta^*_\beta)$ is equal to

$$h_\alpha(u - f(\beta,m)v)^{-1}ww_\beta(-1), \tag{6}$$

providing that $u \neq f(\beta,m)v$. On the other hand, using (4), the second component of $(\eta_\alpha x)\eta^*_\beta$ is equal to

$$w_\alpha(1)wh_\beta(v - f(\beta,m)^{-1}u), \tag{7}$$

providing that $u \neq f(\beta,m)v$. Verification of the identity (6) = (7) in the group W will be left to the intrepid reader.

If $u = f(\beta,m)v$, then the second component of $\eta_\alpha x\eta^*_\beta$ is equal to $w_\alpha(1)ww_\beta(-1)$ for either placement of parentheses!

Case 3. Suppose that $\pi(\beta) = -\alpha$. Using equation (3) we see that the second component of $\eta_\alpha(x\eta^*_\beta)$ is equal to

$$h_\alpha(u - f(-\beta,m)v^{-1})^{-1}wh_\beta(v), \tag{8}$$

providing that $uv \neq f(-\beta,m)$ and that $v \neq 0$. Similarly, using (5), the second component of $(\eta_\alpha x)\eta^*_\beta$ is equal to

$$h_\alpha(u)^{-1}wh_\beta(v - f(-\beta,m)u^{-1}), \tag{9}$$

providing that $u \neq 0$ and $uv \neq f(-\beta,m)$. In order to prove that (8) = (9), we push the factor w to the right in both expressions using 12.5, and substitute $h_{-\alpha}$ for h_α^{-1} in the first factor of each expression. Setting $f = f(-\beta,m)$, it therefore suffices to establish the equation

$$h_{-\alpha}(u - f/v)\,h_{-\alpha}(v)\,c(f^{-1},v) = h_{-\alpha}(u)\,h_{-\alpha}(v - f/u)\,c(f^{-1},v - f/u),$$

or in other words

$$c(u - f/v,v)\,c(f^{-1},v) = c(u,v - f/u)\,c(f^{-1},v - f/u).$$

Substituting $z = f/uv$, and cancelling factors of $c(u,v)$ and $c(f^{-1},v)$ from both sides, it suffices to prove that

$$c(1-z,v) = c(u,1-z)\,c(f^{-1},1-z).$$

But this equality follows easily from the equation $c(z,1-z) = 1$.

REMARK. This is the only place in the entire proof where the full force of the equation $c(z,1-z) = 1$ is needed.

Finally we must see what happens if $u = 0$ or $v = 0$ or $uv = f(-\beta,m)$. In order to take care of these three cases, we must show that

$$h_\alpha(-f(-\beta,m)\,v^{-1})^{-1} w h_\beta(v) = w_\alpha(1)\,w w_\beta(-1),$$

or

$$w_\alpha(1)\,w w_\beta(-1) = h_\alpha(u)^{-1} w h_\beta(-f(-\beta,m)\,u^{-1}),$$

or

$$w_\alpha(1)\,w h_\beta(v) = h_\alpha(u)^{-1} w w_\beta(-1),$$

respectively. The proofs, although tedious, present no particular difficulty. This completes the proof of 12.10. ∎

Proof that the action of G *on* X *is simply transitive* (Lemma 12.8). If $g_1x = g_2x$, then

$$g_1(xg^*) = g_2(xg^*)$$

for every g^* in G^*. But G^* acts transitively, so this proves that $g_1x' = g_2x'$ for every x' in X. Therefore $g_1 = g_2$, which proves 12.8, and completes the proof of 12.9. ∎

Now we are ready to prove Matsumoto's theorem 11.1. Let $c : F^\bullet \times F^\bullet \to A$ be the *universal Steinberg symbol* on the field F. In other words let A be the abelian group which is defined by generators $c(u,v)$ subject only to the relations

$$c(u_1u_2,v) = c(u_1,v)c(u_2,v),\ c(u,v_1v_2) = c(u,v_1)c(u,v_2),$$

and

$$c(u,1-u) = 1,$$

and to the consequences of these relations.

Let $\{u,v\} \in K_2F$ be the symbol of Sections 8 and 9. Since $\{u,v\}$ is bimultiplicative and satisfies $\{u,1-u\} = 1$, there is one and only one homomorphism

$$\eta : A \to K_2F$$

which carries $c(u,v)$ to $\{u,v\}$ for all u and v in F. We must prove that η is an isomorphism.

Let $1 \to A \to G_n \to SL(n,F) \to 1$ be the central extension of Theorem 12.9. It is not difficult to pass to the direct limit as $n \to \infty$, thus obtaining a corresponding central extension

$$1 \to A \to G \to SL(F) \to 1.$$

But according to §5.10 the extension

$$1 \to K_2F \to St(F) \to SL(F) \to 1$$

is the universal central extension of SL(F). So there is one and only one homomorphism

$$\xi : St(F) \to G$$

which covers the identity map of SL(F). Clearly ξ maps K_2F into A. Comparing §8.3 with §12.1, and recalling that $H \cong \lambda(H) \subset G$, we see that ξ carries $\{u,v\}$ to $c(u,v)$ for all u and v. But η carries $c(u,v)$ to $\{u,v\}$ for all u and v. Since K_2F is generated by the symbols $\{u,v\}$, and A is generated by the symbols $c(u,v)$, this completes the proof that $K_2F \cong A$. ∎

§13. More about Dedekind Domains

Let Λ be a Dedekind domain with quotient field F. We will prove

THEOREM 13.1 (Bass, Tate). *There is an exact sequence*

$$K_2F \to \bigoplus K_1\Lambda/\mathfrak{p} \to K_1\Lambda \to K_1F \to \bigoplus K_0\Lambda/\mathfrak{p} \to K_0\Lambda \to K_0F \to 0,$$

where both direct sums extend over all non-zero prime ideals $\mathfrak{p}$ *of* Λ.

Compare Bass, *Algebraic* K-*Theory*, pp. 702, 323. Most of the proof will be given below; but one key step will depend on a reference to Bass.

REMARK. There is of course a natural homomorphism $K_2\Lambda \to K_2F$. If the domain Λ has only countably many ideals, then Bass has recently shown that the extended sequence $K_2\Lambda \to K_2F \to \bigoplus K_1\Lambda/\mathfrak{p}$ is also exact. One would like to push even further to the left. For example, if F is a number field, so that $K_2\Lambda/\mathfrak{p} = 0$, one would conjecture that $K_2\Lambda$ injects into K_2F. But so far this is known only for the special case $\Lambda = Z$. (Compare §10.2 and §8.4.) For a general Dedekind domain, no one has suggested a suitable homomorphism from $\bigoplus K_2\Lambda/\mathfrak{p}$ to $K_2\Lambda$.

To begin the proof of 13.1, let

$$K_0\Lambda/\mathfrak{p} \to K_0\Lambda$$

be the homomorphism which sends the standard generator $[\Lambda/\mathfrak{p}]$ to the difference $[\Lambda^1] - [\mathfrak{p}] \in \tilde{K}_0\Lambda$. Let

$$K_1F \cong F^\bullet \to K_0\Lambda/\mathfrak{p}$$

be the homomorphism which sends each non-zero field element x to the sum

$$\sum v_{\mathfrak{p}}(x)[\Lambda/\mathfrak{p}].$$

Here $v_{\mathfrak{p}}$ denotes the $\mathfrak{p}$-adic valuation of F, and $[\Lambda/\mathfrak{p}]$ denotes the standard generator (= identity element) of $K_0\Lambda/\mathfrak{p}$. Exactness of the resulting sequence

$$K_1\Lambda \to K_1F \to \bigoplus K_0\Lambda/\mathfrak{p} \to K_0\Lambda \to K_0F \to 0$$

is easily verified.

(REMARK. The group $\bigoplus K_0\Lambda/\mathfrak{p}$ can of course be identified with the group of "divisors" of Λ; and is canonically isomorphic to the group of fractional ideals of Λ.)

In order to define the homomorphism

$$K_1\Lambda/\mathfrak{p} \to K_1\Lambda$$

we use a construction due to Mennicke. Let a and b be relatively prime elements of Λ:

$$a\Lambda + b\Lambda = \Lambda .$$

Then there exist elements c and d so that $ad - bc = 1$. Consider the matrix

$$\begin{pmatrix} a & b \\ c & d \end{pmatrix} \in SL(2,\Lambda) \subset GL(\Lambda) .$$

If we work modulo the normal subgroup $E(\Lambda)$, then this matrix depends only on a and b. For if $ad' - bc'$ is also equal to 1, then computation shows that

$$\begin{pmatrix} a & b \\ c & d \end{pmatrix}\begin{pmatrix} a & b \\ c' & d' \end{pmatrix}^{-1} = \begin{pmatrix} 1 & 0 \\ x & 1 \end{pmatrix},$$

for suitable x; so that

$$\begin{pmatrix} a & b \\ c & d \end{pmatrix} \equiv \begin{pmatrix} a & b \\ c' & d' \end{pmatrix} \quad \text{mod } E(\Lambda) .$$

DEFINITION. The Mennicke symbol $\begin{bmatrix} b \\ a \end{bmatrix}$ is defined to be the element of the subgroup $SL(\Lambda)/E(\Lambda) \subset K_1\Lambda$ which is represented by the unimodular matrix $\begin{pmatrix} a & b \\ c & d \end{pmatrix}$.

LEMMA 13.2. *This symbol* $\begin{bmatrix} b \\ a \end{bmatrix} \in K_1\Lambda$, *which is defined whenever* a *and* b *are relatively prime, is symmetric, bimultiplicative, and is not altered if we add a multiple of* a *to* b *or a multiple of* b *to* a.

Here we are thinking of $K_1\Lambda$ as a multiplicative group.

Proof. The properties

$$\begin{bmatrix} b \\ a \end{bmatrix} = \begin{bmatrix} b+\lambda a \\ a \end{bmatrix} \text{ and } \begin{bmatrix} b \\ a \end{bmatrix} = \begin{bmatrix} b \\ a+\lambda b \end{bmatrix}$$

are clear since elementary column operations on a matrix correspond to right multiplication by elementary matrices. It follows that

$$(1) \qquad \begin{bmatrix} b \\ a \end{bmatrix} = \begin{bmatrix} b \\ a+b \end{bmatrix} = \begin{bmatrix} -a \\ a+b \end{bmatrix} = \begin{bmatrix} -a \\ b \end{bmatrix}.$$

Furthermore, if u is a unit of Λ, then

$$(2) \qquad \begin{bmatrix} u \\ a \end{bmatrix} = \begin{bmatrix} u \\ 1 \end{bmatrix} = \begin{bmatrix} 0 \\ 1 \end{bmatrix} = 1 \in K_1\Lambda,$$

since the symbol $\begin{bmatrix} 0 \\ 1 \end{bmatrix}$ clearly corresponds to the identity matrix.

To prove that

$$(3) \qquad \begin{bmatrix} b \\ a \end{bmatrix}\begin{bmatrix} b' \\ a \end{bmatrix} = \begin{bmatrix} bb' \\ a \end{bmatrix}$$

we must verify the congruence

$$\begin{pmatrix} a & b \\ c & d \end{pmatrix}\begin{pmatrix} a & b' \\ c' & d' \end{pmatrix} \equiv \begin{pmatrix} a & bb' \\ * & * \end{pmatrix} \mod E(\Lambda) .$$

But this follows from the matrix identity

$$\begin{pmatrix} 1 & 0 & 0 \\ 0 & 0 & 1 \\ 0 & -1 & 0 \end{pmatrix}\begin{pmatrix} a & b & 0 \\ c & d & 0 \\ 0 & 0 & 1 \end{pmatrix}\begin{pmatrix} 0 & 0 & 1 \\ 1 & 0 & 0 \\ 0 & 1 & 0 \end{pmatrix}\begin{pmatrix} a & b' & 0 \\ c' & d' & 0 \\ 0 & 0 & 1 \end{pmatrix}\begin{pmatrix} 0 & 0 & -1 \\ 0 & 1 & 0 \\ 1 & 0 & b \end{pmatrix}$$

$$= \begin{pmatrix} a & bb' & 0 \\ * & * & * \\ * & * & 1 \end{pmatrix} \equiv \begin{pmatrix} a & bb' & 0 \\ * & * & 0 \\ 0 & 0 & 1 \end{pmatrix} \mod E(\Lambda) ;$$

since it is easily checked that the first, third, and fifth matrices in the five-fold product belong to $E(\Lambda)$.

Now combining (1), (2), and (3), we have

$$(4) \qquad \begin{bmatrix} b \\ a \end{bmatrix} = \begin{bmatrix} -a \\ b \end{bmatrix} = \begin{bmatrix} -1 \\ b \end{bmatrix}\begin{bmatrix} a \\ b \end{bmatrix} = \begin{bmatrix} a \\ b \end{bmatrix}.$$

Since the bimultiplicative property follows from (3) and (4), this proves Lemma 13.2. ∎

Now consider a non-zero ideal $\mathfrak{a} \subset \Lambda$ and a ring element b which is relatively prime to $\mathfrak{a}$.

LEMMA 13.3. *There is one and only one bimultiplicative symbol*

$$\begin{bmatrix} b \\ \mathfrak{a} \end{bmatrix} \in SL(\Lambda)/E(\Lambda) ,$$

which is defined whenever b *and the non-zero ideal* $\mathfrak{a}$ *are relatively prime, which depends only on the residue class of* b *modulo* $\mathfrak{a}$, *and which coincides with the Mennicke symbol* $\begin{bmatrix} b \\ a \end{bmatrix}$ *when* $\mathfrak{a}$ *is the principal ideal generated by* a.

Proof. (Compare Bass, *Algebraic* K-*Theory*, p. 308.) Given any non-zero ideal $\mathfrak{a}$ we can choose a relatively prime ideal $\mathfrak{x}$ belonging to the ideal class $\{\mathfrak{a}\}^{-1}$. (Compare §1.8.) The product $\mathfrak{a}\mathfrak{x}$ will then be a principal ideal, generated say by c. By the Chinese remainder theorem, there exists an element b′ of Λ so that

$$b' \equiv b \bmod \mathfrak{a}, \quad b' \equiv 1 \bmod \mathfrak{x}.$$

Now suppose that a symbol $\begin{bmatrix} b \\ \mathfrak{a} \end{bmatrix}$, satisfying the conditions above, is given. Then $\begin{bmatrix} b \\ \mathfrak{a} \end{bmatrix}$ must be equal to

$$\begin{bmatrix} b \\ \mathfrak{a} \end{bmatrix}\begin{bmatrix} 1 \\ \mathfrak{x} \end{bmatrix} = \begin{bmatrix} b' \\ \mathfrak{a} \end{bmatrix}\begin{bmatrix} b' \\ \mathfrak{x} \end{bmatrix} = \begin{bmatrix} b' \\ \mathfrak{a}\mathfrak{x} \end{bmatrix} = \begin{bmatrix} b' \\ c \end{bmatrix}.$$

This proves uniqueness.

To prove existence, we define $\begin{bmatrix} b \\ \mathfrak{a} \end{bmatrix}$ to be the element $\begin{bmatrix} b' \\ c \end{bmatrix}$ of $SK_1(\Lambda)$, where b′ and c are constructed as above. This definition does not depend on the choice of b′, since if b″ also satisfies

$$b'' \equiv b \bmod \mathfrak{a}, \quad b'' \equiv 1 \bmod \mathfrak{x},$$

then $b'' \equiv b' \pmod c$, hence $\begin{bmatrix} b'' \\ c \end{bmatrix} = \begin{bmatrix} b' \\ c \end{bmatrix}$. Further this definition does not depend on the choice of c. For if uc is another generator for the ideal $\mathfrak{a}\mathfrak{x}$, then $\begin{bmatrix} b' \\ uc \end{bmatrix} = \begin{bmatrix} b' \\ c \end{bmatrix}$ by (2) above. Finally, the definition does not depend on the choice of $\mathfrak{x}$. For if $\mathfrak{x}'$ is another ideal prime to $\mathfrak{a}$, with $\mathfrak{a}\mathfrak{x}' = \Lambda c'$, then choosing $\mathfrak{y}$ in the ideal class of $\mathfrak{a}$ but prime to $\mathfrak{a}$ we have

$$\mathfrak{x}\mathfrak{y} = \Lambda d, \quad \mathfrak{x}'\mathfrak{y} = \Lambda d'$$

for some elements d and d′. Note that the ratio $(cd')/(c'd)$ is a unit of Λ, since both cd' and $c'd$ generate the ideal $\mathfrak{a}\mathfrak{x}\mathfrak{x}'\mathfrak{y}$. Now choosing b′ so that

$$b' \equiv b \bmod \mathfrak{a}, \quad b' \equiv 1 \bmod \mathfrak{x}\mathfrak{x}'\mathfrak{y},$$

we have $b' \equiv 1 \pmod{d}$ and $b' \equiv 1 \pmod{d'}$, so that

$$\begin{bmatrix} b' \\ c \end{bmatrix} = \begin{bmatrix} b' \\ cd' \end{bmatrix} = \begin{bmatrix} b' \\ c'd \end{bmatrix} = \begin{bmatrix} b' \\ c' \end{bmatrix}$$

as required.

It is now easy to check that this symbol $\begin{bmatrix} b \\ \mathfrak{a} \end{bmatrix}$ is bimultiplicative, depends only on the residue class of b mod $\mathfrak{a}$, and coincides with $\begin{bmatrix} b \\ a \end{bmatrix}$ when $\mathfrak{a} = \Lambda a$. This completes the proof of 13.3. ■

REMARK. If u is a unit of Λ, this generalized Mennicke symbol $\begin{bmatrix} u \\ \mathfrak{a} \end{bmatrix}$ need not be trivial. (Compare 13.5.) It can be described as a product as follows. Let (u) denote the element of $K_1\Lambda$ corresponding to u.

ASSERTION 13.4. *The symbol* $\begin{bmatrix} u \\ \mathfrak{a} \end{bmatrix} \in K_1\Lambda$ *is equal to the product of* $(u) \in K_1\Lambda$ *with the element* $1 - [\mathfrak{a}] \in K_0\Lambda$.

In other words the composition $K_1\Lambda \to K_1\Lambda/\mathfrak{a} \to K_1\Lambda$ is just multiplication by $1 - [\mathfrak{a}]$. (Compare the behavior of the "transfer" in §14.1.)

Proof. Choosing an ideal $\mathfrak{x}$ with $\mathfrak{a} + \mathfrak{x} = \Lambda$ and $\mathfrak{a}\mathfrak{x} = \Lambda c$, it follows from the exact sequence

$$0 \to \Lambda \xrightarrow{(c,-c)} \mathfrak{a} \oplus \mathfrak{x} \to \Lambda \to 0$$

that the direct sum $\mathfrak{a} \oplus \mathfrak{x}$ is free. To compute the product $(u)[\mathfrak{a}] \in K_1\Lambda$ we must study the automorphism $T : \mathfrak{a} \oplus \mathfrak{x} \to \mathfrak{a} \oplus \mathfrak{x}$ which is the identity on the second summand, and multiplies each element of the first summand by u. (Compare p. 27.)

Choose $a \in \mathfrak{a}$ so that $1 - a \in \mathfrak{r}$. As basis for $\mathfrak{a} \oplus \mathfrak{r}$ we will use the two vectors

$$b_1 = (a, 1-a), \quad b_2 = (c, -c).$$

Computation shows that

$$T(b_2) = c(u-1)\,b_1 + (u+a-ua)\,b_2,$$

so that the matrix of T has the form

$$\begin{pmatrix} * & c(u-1) \\ * & u+a-ua \end{pmatrix} \in GL(2, \Lambda).$$

To compute $(u)(1 - [\mathfrak{a}])$ we must pass to the matrix

$$\begin{pmatrix} 1 & 0 \\ 0 & u \end{pmatrix} \begin{pmatrix} * & c(u-1) \\ * & u+a-ua \end{pmatrix}^{-1} = \begin{pmatrix} u+a-ua & c(1-u) \\ * & * \end{pmatrix}$$

in $SL(2, \Lambda)$. The corresponding Mennicke symbol is evidently

$$\begin{bmatrix} c(1-u) \\ u+a-ua \end{bmatrix} = \begin{bmatrix} u+a-ua \\ \mathfrak{a} \end{bmatrix} \begin{bmatrix} u+a-ua \\ \mathfrak{r}(1-u) \end{bmatrix}.$$

But

$$u+a-ua \equiv u \bmod \mathfrak{a}, \qquad u+a-ua \equiv 1 \bmod \mathfrak{r}(1-u),$$

so this product is equal to $\begin{bmatrix} u \\ \mathfrak{a} \end{bmatrix}$, as required. ∎

Here is an example to illustrate these constructions.

EXAMPLE 13.5. Let Λ be obtained from the real polynomial ring $R[x]$ by adjoining an element y satisfying the equation $x^2 + y^2 = 1$. *Then the matrix*

$$\begin{pmatrix} x & y \\ -y & x \end{pmatrix} \in SL(2, \Lambda)$$

represents a non-trivial element of $K_1\Lambda$. This will be proved using the techniques of §7. Note that Λ can be identified with the ring consisting of all real polynomial functions on the unit circle. Hence every matrix in $SL(n, \Lambda)$ gives rise to a polynomial mapping from the unit circle to $SL(n, R)$. Since elementary matrices give rise to null-homotopic maps, this construction defines a homomorphism $K_1\Lambda \to \pi_1 SL(R) \cong Z/2Z$. But the matrix $\begin{pmatrix} x & y \\ -y & x \end{pmatrix}$ corresponds to a generator of $\pi_1 SL(2, R)$, and hence maps to the non-trivial element of $\pi_1 SL(R)$.

Thus the Mennicke symbol $\begin{bmatrix} y \\ x \end{bmatrix}$ is non-trivial. Since the principal ideal Λx splits as a product $\mathfrak{p}\mathfrak{q}$, where

$$y \equiv 1 \bmod \mathfrak{p}, \quad y \equiv -1 \bmod \mathfrak{q},$$

it follows that

$$1 \neq \begin{bmatrix} y \\ x \end{bmatrix} = \begin{bmatrix} y \\ \mathfrak{p} \end{bmatrix}\begin{bmatrix} y \\ \mathfrak{q} \end{bmatrix} = \begin{bmatrix} -1 \\ \mathfrak{q} \end{bmatrix}.$$

(Here $\mathfrak{p}$ [respectively $\mathfrak{q}$] is the ideal consisting of polynomial functions which vanish at the point $(0,1)$ [respectively $(0,-1)$] of the unit circle.) Therefore the element

$$\begin{bmatrix} -1 \\ \mathfrak{q} \end{bmatrix} = (-1)(1-[\mathfrak{q}]) \in SK_1\Lambda$$

is non-trivial. (In fact it is not difficult to show that $SK_1\Lambda$ is cyclic of order 2 with generator $\begin{bmatrix} -1 \\ \mathfrak{q} \end{bmatrix}$. Compare Bass, *Algebraic* K-*Theory*, p. 714.)

If $R(x,y)$ denotes the quotient field of Λ, it is amusing to speculate about the kernel of the natural homomorphism $K_2\Lambda \to K_2R(x,y)$, and as to whether it equals the image of a suitable homomorphism from $\bigoplus K_2\Lambda/\mathfrak{q}$. If such a homomorphism exists, then the composition $K_2\Lambda \to K_2\Lambda/\mathfrak{q} \to K_2\Lambda$ should be multiplication by $1-[\mathfrak{q}]$. Hence the element $\{-1,-1\} \in K_2\Lambda/\mathfrak{q}$ should map to the product

$$\{-1,-1\}(1-[\mathfrak{q}]) = (-1)^2(1-[\mathfrak{q}]) = (-1)\begin{bmatrix} y \\ x \end{bmatrix}$$

in the kernel of $K_2\Lambda \to K_2R(x,y)$. The question as to whether or not $(-1)\begin{bmatrix} y \\ x \end{bmatrix}$ is trivial seems very difficult.

We return to the proof of Theorem 13.1. Let Λ be any Dedekind domain, and let $\mathfrak{p}$ be a non-zero prime ideal. Then the correspondence

$$b \mapsto \begin{bmatrix} b \\ \mathfrak{p} \end{bmatrix}$$

gives rise to a homomorphism from the group $(\Lambda/\mathfrak{p})^{\bullet} = K_1\Lambda/\mathfrak{p}$ to $K_1\Lambda$. Forming the direct sum over all non-zero primes, we obtain the required homomorphism $\bigoplus K_1\Lambda/\mathfrak{p} \to K_1\Lambda$.

In order to prove that the resulting sequence

$$\bigoplus K_1\Lambda/\mathfrak{p} \to K_1\Lambda \to K_1F$$

is exact, we will need the following:

LEMMA 13.6 (Bass). *Every matrix in* $SL(\Lambda)$ *can be reduced by elementary row and column operations to a matrix in the subgroup* $SL(2,\Lambda)$.

The proof works for any commutative Λ satisfying the condition that a non-zero element is contained in only finitely many maximal ideals.

Proof of 13.6. Let $(a_1,\ldots,a_n)$ be the last row of an arbitrary matrix A in $SL(n,\Lambda)$, with $n \geq 3$. Clearly

$$\Lambda a_1 + \ldots + \Lambda a_n = \Lambda .$$

Case 1. If the ideal generated by $a_1,a_2,\ldots,a_{n-1}$ is already equal to the entire ring Λ, then elementary column operations can be used to replace a_n by 1. Suitable row and column operations will then replace the matrix A by a matrix of the form $\mathrm{diag}(A',1)$, lying in the subgroup $SL(n-1,\Lambda)$.

Case 2. If $a_2 = 0$, then elementary column operations will replace a_2 by 1, and we can proceed as in Case 1.

Case 3. If $a_2 \neq 0$, then there can only be finitely many maximal ideals, say $\mathfrak{m}_1,\ldots,\mathfrak{m}_s$, which contain the elements $a_2,a_3,\ldots,$ and a_{n-1}. Of these ideals, suppose that the first r contain a_1, but that the remaining $s-r$ do not contain a_1. Choose an element e of Λ so that

$$e \equiv 1 \mod \mathfrak{m}_1, \mathfrak{m}_2,\ldots, \mathfrak{m}_r,$$
$$e \equiv 0 \mod \mathfrak{m}_{r+1},\ldots, \mathfrak{m}_s.$$

Now adding e times the last column to the first, we replace a_1 by $a_1 + ea_n$. Clearly the ideal generated by the elements

$$a_1+ea_n,\ a_2,\ a_3,\ \ldots,\ a_{n-1}$$

is equal to the entire ring Λ. Hence we can proceed as in Case 1.

Thus the given matrix is congruent modulo $E(\Lambda)$ to a matrix in $SL(n-1,\Lambda)$, and continuing inductively, it is congruent to a matrix in $SL(2,\Lambda)$. This proves 13.6. ∎

Proof of exactness of the sequence $\bigoplus K_1\Lambda/\mathfrak{p} \to K_1\Lambda \to K_1F$. The image of the first homomorphism is clearly the subgroup of $K_1\Lambda$ generated by all Mennicke symbols $\begin{bmatrix} b \\ a \end{bmatrix}$, and hence is equal to the image of $SL(2,\Lambda)$ in $K_1\Lambda$. On the other hand the kernel of the second homomorphism is clearly the subgroup $SL(\Lambda)/E(\Lambda)$ of $K_1\Lambda$. But by 13.6 the image of $SL(2,\Lambda)$ is precisely equal to $SL(\Lambda)/E(\Lambda)$. This proves that the above sequence is exact.

Next we will prove the following. Again let Λ be a Dedekind domain with quotient field F.

LEMMA 13.7 (Tate). *The group* K_2F *is generated by those symbols* $\{a, b\}$ *for which* a *and* b *are relatively prime elements of the domain* Λ.

Proof. Let L be the subgroup of K_2F generated by all symbols $\{a, b\}$, with a and b relatively prime in Λ. For any x and y in $F^\bullet$ we must prove that $\{x, y\} \in L$. The proof will be by induction on the number of maximal ideals $\mathfrak{p}$ for which both $v_\mathfrak{p}(x) \neq 0$ and $v_\mathfrak{p}(y) \neq 0$ (where $v_\mathfrak{p}$ is the $\mathfrak{p}$-adic valuation).

Case 1. Suppose there are *no* primes $\mathfrak{p}$ with $v_\mathfrak{p}(x)\, v_\mathfrak{p}(y) \neq 0$. In the group of fractional ideals of Λ, we can express the fractional ideal Λx uniquely as the quotient $\mathfrak{a}/\mathfrak{b}$ of two relatively prime ideals contained in Λ. Similarly we can write $\Lambda y = \mathfrak{c}/\mathfrak{d}$; where $\mathfrak{a}\mathfrak{b}$ is prime to $\mathfrak{c}\mathfrak{d}$ by hypothesis.

According to Lemma 1.8, there exists an ideal $\mathfrak{e}$ which belongs to the ideal class of $\mathfrak{b}^{-1}$, and is relatively prime to $\mathfrak{c}\mathfrak{d}$. Setting

$$\mathfrak{b}\mathfrak{e} = \Lambda b, \quad \text{and} \quad a = bx,$$

we have expressed x as a quotient $x = a/b$ with a and b relatively prime to $\mathfrak{c}\mathfrak{d}$. Similarly we can write y as a quotient c/d with c and d relatively prime to both a and b. Then

$$\{x, y\} = \{a, c\}\{b, d\}\{a, d\}^{-1}\{b, c\}^{-1},$$

which proves that $\{x, y\} \in L$.

Case 2. Suppose that there is just one prime $\mathfrak{p}$ with $v_{\mathfrak{p}}(x)\, v_{\mathfrak{p}}(y) \neq 0$. Choose an element z of $F^{\bullet}$ so that

$$v_{\mathfrak{p}}(z) = -1,$$

but so that

$$v_{\mathfrak{q}}(z) \geq 0 \quad \text{and} \quad v_{\mathfrak{q}}(z)\, v_{\mathfrak{q}}(y) = 0$$

for every prime $\mathfrak{q} \neq \mathfrak{p}$. (In fact, let $\Lambda z = \mathfrak{p}^{-1}\mathfrak{r}$ where $\mathfrak{r} \in \{\mathfrak{p}\}$ is prime to every $\mathfrak{q}$ with $v_{\mathfrak{q}}(y) \neq 0$.) Setting

$$i = v_{\mathfrak{p}}(x), \quad j = v_{\mathfrak{p}}(y),$$

note that $\{z^i x, y\} \in L$ by Case 1. Hence

$$\{x, y\} \equiv \{z, y\}^{-i} \bmod L. \tag{1}$$

But a short argument shows that the element $y(1-z)^j$ satisfies the condition

$$v_{\mathfrak{q}}(z)\, v_{\mathfrak{q}}(y(1-z)^j) = 0$$

for *every* prime $\mathfrak{q}$. Therefore

$$\{z, y(1-z)^j\} \in L,$$

also by Case 1. Since $\{z, 1-z\} = 1$, this proves that $\{z, y\} \in L$. Using the congruence (1), it follows that $\{x, y\} \in L$.

Case 3. Finally suppose that there are n distinct primes $\mathfrak{p}_1, \ldots, \mathfrak{p}_n$ for which

$$v_{\mathfrak{p}_i}(x)\, v_{\mathfrak{p}_i}(y) \neq 0;$$

with $n > 1$. Choose an element w of $F^{\bullet}$ so that

$$v_{\mathfrak{p}_1}(w) = -v_{\mathfrak{p}_1}(x),$$

but so that

$$v_{\mathfrak{q}}(w)\, v_{\mathfrak{q}}(y) = 0 \quad \text{for } \mathfrak{q} \neq \mathfrak{p}_1.$$

Then

$$\{xw, y\} \in L \quad \text{and} \quad \{w, y\} \in L$$

by the induction hypothesis; so it follows again that $\{x, y\} \in L$. This completes the proof of 13.7. ∎

Now define a homomorphism

$$d : K_2F \to \bigoplus K_1\Lambda/\mathfrak{p}$$

as follows. Each generator $\{x, y\}$ of K_2F is to map to the element in the direct sum whose $\mathfrak{p}$-th coordinate is the tame symbol $d_{\mathfrak{p}}(x, y)$ of §11.5. We will prove the following.

LEMMA 13.8. *The composition of the homomorphisms*

$$K_2F \xrightarrow{d} \bigoplus K_1\Lambda/\mathfrak{p} \to K_1\Lambda$$

is zero.

Proof. By 13.7, it suffices to prove that the element $\{a, b\}$ in K_2F maps to zero in $K_1\Lambda$ whenever a and b are relatively prime elements of Λ. Setting

$$\Lambda a = \mathfrak{p}_1^{m_1} \cdots \mathfrak{p}_r^{m_r} \text{ and } \Lambda b = \mathfrak{q}_1^{n_1} \cdots \mathfrak{q}_s^{n_s},$$

we see that the image of $\{a, b\}$ in $K_1\Lambda/\mathfrak{q}_j$ is equal to the residue class of a^{n_j} modulo $\mathfrak{q}_j$. Similarly the image of $\{a, b\}$ in $K_1\Lambda/p_i$ is the reciprocal of the residue class of b^{m_i} modulo $\mathfrak{p}_i$. Therefore the image of $\{a, b\}$ in $K_1\Lambda$ is equal to the quotient of

$$\begin{bmatrix} a^{n_1} \\ \mathfrak{q}_1 \end{bmatrix} \cdots \begin{bmatrix} a^{n_s} \\ \mathfrak{q}_s \end{bmatrix} = \begin{bmatrix} a \\ \mathfrak{q}_1^{n_1} \end{bmatrix} \cdots \begin{bmatrix} a \\ \mathfrak{q}_s^{n_s} \end{bmatrix} = \begin{bmatrix} a \\ b \end{bmatrix}$$

by

$$\begin{bmatrix} b^{m_1} \\ \mathfrak{p}_1 \end{bmatrix} \cdots \begin{bmatrix} b^{m_r} \\ \mathfrak{p}_r \end{bmatrix} = \begin{bmatrix} b \\ \mathfrak{p}_1^{m_1} \end{bmatrix} \cdots \begin{bmatrix} b \\ \mathfrak{p}_r^{m_r} \end{bmatrix} = \begin{bmatrix} b \\ a \end{bmatrix}.$$

Since $\begin{bmatrix} a \\ b \end{bmatrix} = \begin{bmatrix} b \\ a \end{bmatrix}$ by 13.2, this completes the proof of Lemma 13.8. ∎

The proof that the sequence of 13.8 is actually exact will be based on a very pretty result which we state without proof. First a definition. Our previous concept of "Mennicke symbol" can be generalized as follows. Let $W_\Lambda \subset \Lambda \times \Lambda$ be the set consisting of all pairs (x, y) of relatively prime elements of Λ.

DEFINITION. A *Mennicke function* on Λ with values in a commutative group C will mean a bimultiplicative function

$$\mu : W_\Lambda \to C$$

such that $\mu(x, y)$ is unchanged if we add a multiple of x to y or a multiple of y to x.

Let N denote the kernel of the natural surjection $GL(2, \Lambda) \to K_1\Lambda$. (Compare 13.6.)

KUBOTA-BASS THEOREM. *If* $\mu : W_\Lambda \to C$ *is a Mennicke function on the Dedekind domain* Λ, *then the correspondence*

$$\begin{pmatrix} a & b \\ c & d \end{pmatrix} \mapsto \mu(a, b)$$

defines a homomorphism from $GL(2, \Lambda)$ *to* C *which annihilates* N.

Hence there exists one and only one homomorphism $\bar{\mu}$ from $K_1\Lambda \cong GL(2, \Lambda)/N$ to C so that the following diagram commutes:

$$\begin{array}{ccc} GL(2, \Lambda) & \xrightarrow{\text{first row}} & W_\Lambda \\ \downarrow & & \downarrow \mu \\ K_1\Lambda & \xrightarrow{\bar{\mu}} & C \end{array}$$

For the proof of this theorem, we refer the reader to Bass, *Algebraic K-Theory*, p. 298, or to Bass, Milnor, and Serre, Publ. Math. I.H.E.S. 33, §§6-9.

The Kubota-Bass Theorem will be applied as follows. Let

$$C = (\bigoplus K_1\Lambda/\mathfrak{p})/d(K_2F)$$

be the cokernel of the homomorphism d. Then we will construct a Mennicke function on Λ with values in C. For each $(a, b) \in W_\Lambda$, let

$$\psi(a, b) \in \bigoplus K_1\Lambda/\mathfrak{p}$$

be the element whose $\mathfrak{p}$-th coordinate is equal to

$$(b \bmod \mathfrak{p})^{v_{\mathfrak{p}}(a)}$$

if $\mathfrak{p}$ divides a, and to 1 otherwise. Evidently the function ψ is bi-multiplicative, and $\psi(a, b)$ is unchanged if we add a multiple of a to b. Furthermore note the congruence

$$\psi(a, b) \equiv \psi(b, a) \mod d(K_2F).$$

In fact direct computation shows that

$$\psi(b, a)/\psi(a, b) = d\{a, b\}.$$

Therefore, defining $\mu(a, b) \in C$ as the residue class of $\psi(a, b)$ modulo $d(K_2F)$, it follows that $\mu : W_\Lambda \to C$ is a Mennicke function. Thus, according to Kubota and Bass, there is one and only one homomorphism $\bar{\mu} : K_1\Lambda \to C$ which carries the class of $\begin{pmatrix} a & b \\ c & d \end{pmatrix}$ to $\mu(a, b)$.

Next note that the composition $\bigoplus K_1\Lambda/\mathfrak{p} \to K_1\Lambda \overset{\bar{\mu}}{\to} C$ is equal to the natural projection homomorphism (reduction modulo $d(K_2F)$. In fact the image of a given element (b mod $\mathfrak{p}$) in the $\mathfrak{p}$-th summand can be evaluated as follows. By definition the image $\begin{bmatrix} b \\ \mathfrak{p} \end{bmatrix}$ in $K_1\Lambda$ is equal to $\begin{bmatrix} b' \\ a \end{bmatrix}$ where

$$\mathfrak{p}\,\mathfrak{x} = \Lambda a, \quad b' \equiv b \bmod \mathfrak{p}, \quad b' \equiv 1 \bmod \mathfrak{x}$$

for some ideal $\mathfrak{x}$. Then

$$\bar{\mu}\begin{bmatrix} b \\ \mathfrak{p} \end{bmatrix} = \bar{\mu}\begin{bmatrix} b' \\ a \end{bmatrix} = \mu(a, b') = (\psi(a, b') \bmod d(K_2F)).$$

Inspection shows that the $\mathfrak{p}$-th component of $\psi(a, b')$ is equal to (b mod $\mathfrak{p}$), and that all other components are trivial.

Thus if an element in $\bigoplus K_1\Lambda/\mathfrak{p}$ maps to the identity in $K_1\Lambda$, and hence maps to the identity in C, then it must belong to $d(K_2F)$. This shows that the sequence of 13.8 is exact, and completes the proof of 13.1. ∎

§14. The Transfer Homomorphism

The results of this section are largely due to Bass and Tate.

Let Λ be a ring (always associative with 1), and let $\Gamma \supset \Lambda$ be a larger ring such that Γ is finitely generated and projective when considered as left Λ-module. The inclusion homomorphism will be denoted by $f : \Lambda \to \Gamma$. We assume of course that $f(1) = 1$.

This section will define the *transfer homomorphism*

$$f^* : K_i\Gamma \to K_i\Lambda$$

for $i = 0, 1, 2,$ and prove the following property.

THEOREM 14.1. *If* Γ *is commutative, and* $i, j \leq 2$, *then the identity*

$$f^*(x \cdot f_*(y)) = (f^*x) \cdot y$$

is valid for every $x \in K_i\Gamma$ *and* $y \in K_j\Lambda$.

In other words the transfer homomorphism to $K_*\Lambda$ is $K_*\Lambda$-linear. As an example, taking x to be the identity element $1 = [\Gamma] \in K_0\Gamma$, we obtain the formula

$$f^*(f_*(y)) = f^*(1) \cdot y .$$

(Compare §13.4.)

REMARK 1. Bass defines the transfer homomorphisms $K_0\Gamma \to K_0\Lambda$ and $K_1\Gamma \to K_1\Lambda$ assuming only that there exists a finite Λ-linear resolution

$$0 \to P_n \to P_{n-1} \to \dots \to P_0 \to \Gamma \to 0 ,$$

where the P_k are finitely generated and projective over Λ. (*Algebraic K-Theory*, p. 451.) In this generality, the transfer homomorphism includes the homomorphisms $K_0\Lambda/\mathfrak{a} \to K_0\Lambda$ and $K_1\Lambda/\mathfrak{a} \to K_1\Lambda$ of §13. It would be of great interest to know whether or not the transfer homomorphism for K_2 can also be defined in this more general context.

REMARK 2. The terms "norm" or "restriction of scalars" are both sometimes used for the transfer homomorphism. Certainly both terms are quite descriptive. I feel that the term transfer is useful since it can also be used for related homomorphisms in homological algebra and in topology. Thus, if Π is a group and Π' a subgroup of finite index, then the classical transfer homomorphism

$$H_i(\Pi) \to H_i(\Pi')$$

is related to our

$$f^* : K_i(Z\Pi) \to K_i(Z\Pi');$$

and is a special case of the "topological transfer homomorphism" from the homology of a space B to the homology of a finite covering space E. The analogous homomorphism $K^*(E) \to K^*(B)$ of topological K-theory is closely related to our transfer $K_*(C^E) \to K_*(C^B)$, where C^E denotes the ring of continuous complex valued functions on E. (Compare §7.)

Definition of the transfer. Every finitely generated projective P over Γ can also be considered as a finitely generated projective over the subring Λ. We will denote the resulting Λ-module by P_Λ, whenever it is necessary to make the distinction.

There is clearly one and only one homomorphism

$$f^*: K_0\Gamma \to K_0\Lambda$$

which carries each generator $[P]$ of $K_0\Gamma$ to the generator $[P_\Lambda]$ of $K_0\Lambda$.

To define the transfer on K_1 and K_2, we first define an embedding

$$f^\# : GL(n,\Gamma) \to GL(\Lambda).$$

Any matrix $X \in GL(n,\Gamma)$ gives rise to a Γ-linear automorphism of the free module Γ^n, which will be denoted by X also. Choose a projective Q over Λ so that the direct sum $\Gamma \oplus Q$ is free, say on r generators, over Λ. Then $\Gamma^n \oplus Q^n$ is also Λ-free, and we can consider the Λ-linear automorphism

$$X \oplus (\text{identity map of } Q^n)$$

of $\Gamma^n \oplus Q^n$. Choosing a Λ-basis for this direct sum, the automorphism $X \oplus (\text{identity})$ is represented by a matrix which we denote by

$$f^{\#}(X) \in GL(nr,\Lambda) \subset GL(\Lambda)\,.$$

Proceeding as in §3.2, this homomorphism $f^{\#}$ is well defined up to inner automorphism of $GL(\Lambda)$.

Now abelianizing, and taking the direct limit as $n \to \infty$, we obtain the transfer homomorphism from $K_1\Gamma$ to $K_1\Lambda$. Similarly, taking the Schur multiplier of the induced homomorphism

$$E(n,\Gamma) \to E(\Lambda),$$

and then passing to the limit as $n \to \infty$, we obtain the transfer from $K_2\Gamma$ to $K_2\Lambda$. Details will be left to the reader.

DEFINITION 14.2. If Λ is commutative, then composing the natural homomorphisms

$$\Gamma^{\bullet} \to K_1\Gamma \xrightarrow{f^*} K_1\Lambda \xrightarrow{\det} \Lambda^{\bullet}$$

we obtain the *norm homomorphism* norm: $\Gamma^{\bullet} \to \Lambda^{\bullet}$. As an example, if Γ is free over Λ with basis $b_1, \ldots, b_n$, then setting $\gamma b_i = \sum \lambda_{ij}\, b_j$ the norm of γ is clearly equal to the determinant of the matrix (λ_{ij}). Compare van der Waerden, *Modern Algebra* I.

The proof of Theorem 14.1 will be based on the following associative law. Let M be a Γ-module and N a Λ-module, with $\Gamma \supset \Lambda$ commutative. Then

$$M \underset{\Lambda}{\otimes} N = (M \underset{\Gamma}{\otimes} \Gamma) \underset{\Lambda}{\otimes} N$$

is canonically isomorphic to $M \underset{\Gamma}{\otimes} (\Gamma \underset{\Lambda}{\otimes} N)$.

To simplify the notation, we will denote either tensor product simply by $M \otimes N$. Thus the symbol $\otimes$ stands for the tensor product over Λ; but we note that $M \otimes N$ has the structure of a Γ-module whenever M is a Γ-module.

Proof of 14.1 *for the case* $i = j = 1$. Given

$$x \in K_1\Gamma, \quad y \in K_1\Lambda.$$

choose a representative $X \in GL(m,\Gamma)$ for x, and a representative $Y \in GL(n,\Lambda)$ for y. We will think of X and Y as automorphisms of the free modules Γ^m and Λ^n respectively. Let I and I′ denote the corresponding identity automorphisms. Form the direct sum of three copies of $\Gamma^m \otimes \Lambda^n$, and consider the Γ-linear automorphisms

$$\xi = (X \otimes I') \oplus (X \otimes I')^{-1} \oplus (I \otimes I')$$

and

$$\eta = (I \otimes Y) \oplus (I \otimes I') \oplus (I \otimes Y)^{-1}$$

of this direct sum. Lifting ξ and η to the Steinberg group, we can form the commutator

$$\xi \star \eta \in K_2\Gamma.$$

By definition, $\xi \star \eta$ is equal to the product $x \cdot f_* y$.

In order to transfer to $K_2\Lambda$, it evidently suffices to apply the homomorphism

$$f^\# : GL(3mn, \Gamma) \to GL(\Lambda)$$

to ξ and η. Choose Q so that $\Gamma \oplus Q$ is Λ-free, and let ι denote the identity map of Q^{3mn}. Thus $f^*(x \cdot f_* y)$ is equal to the commutator

$$(\xi \oplus \iota) \star (\eta \oplus \iota) \in K_2\Lambda .$$

Now we must compute $f^*(x) \cdot y$. The element $f^*(x) \in K_1\Lambda$ is represented by the Λ-linear automorphism $X \oplus I''$ of the direct sum $\Gamma^m \oplus Q^m$; where I'' denotes the identity map of Q^m. Form the direct sum of three copies of $(\Gamma^m \oplus Q^m) \otimes \Lambda^n$. Setting

$$\xi_1 = (X \oplus I'') \otimes I' \oplus (X \oplus I'')^{-1} \otimes I' \oplus (I \oplus I'') \otimes I'$$

and

$$\eta_1 = (I \oplus I'') \otimes Y \oplus (I \oplus I'') \otimes I' \oplus (I \oplus I'') \otimes Y^{-1},$$

it follows that

$$f^*(x) \cdot y = \xi_1 \star \eta_1 \in K_2\Lambda .$$

In order to transform these expressions into something more amenable, we replace each $(\Gamma^m \oplus Q^m) \otimes \Lambda^n$ by $(\Gamma^m \otimes \Lambda^n) \oplus Q^{mn}$, and then shift all of the Q^{mn} summands to the right. Thus ξ_1 is transformed into $\xi \oplus \iota$, and η_1 is transformed into something of the form $\eta \oplus \zeta$. Hence

$$f^*(x) \cdot y = (\xi \oplus \iota) \star (\eta \oplus \zeta) \in K_2\Gamma .$$

Since the commutator is bimultiplicative (§8.1), this expression is equal to the product of

$$(\xi\oplus\iota)\star(\eta\oplus\iota) = f^*(x\cdot f_*y)$$

and of the element

$$(\xi\oplus\iota)\star(1\oplus\zeta)$$

which is clearly equal to 1.

This completes the proof when $i = j = 1$. The remaining cases of 14.1 will be left to the reader. ∎

An Application to Fields

Let F be a field, let F[x] be the ring of polynomials in one indeterminate, and let α and β be elements of $F^\bullet \cong K_1F$. For any positive integer n, we can consider the principal ideal

$$(x^n-\beta) \subset F[x],$$

and the corresponding quotient ring

$$\Gamma = F[x]/(x^n-\beta),$$

which has dimension n over F.

REMARK. If the polynomial $x^n-\beta$ is irreducible, then Γ is a field. More generally if $x^n-\beta$ splits into k distinct irreducible factors, then Γ splits as the cartesian product of k extension fields, each of which contains one or more n-th roots of β.

COROLLARY 14.3. *If* α *belongs to the image of the norm homomorphism*

$$\text{norm}: \Gamma^\bullet \to F^\bullet,$$

where $\Gamma = F[x]/(x^n-\beta)$, *then the element*

$$\{\alpha,\beta\} \in K_2F$$

has an n-*th root in (the multiplicative group)* K_2F.

The conclusion of 14.3 can be expressed briefly by the congruence

$$\{\alpha,\beta\} \equiv 1 \bmod (K_2F)^n,$$

where $(K_2F)^n$ denotes the subgroup consisting of all n-th powers in K_2F.

Proof. Recall from 14.2 that the norm homomorphism

$$\text{norm}: \Gamma^{\bullet} \to F^{\bullet}$$

can be identified with the transfer homomorphism $K_1\Gamma \to K_1F$. (The group $K_1\Gamma$ is clearly isomorphic to $\Gamma^{\bullet}$, since Γ is a cartesian product of fields or of local rings.) Thus 14.1 can be written in the form

$$f^*\{\gamma, f(\beta)\} = \{\text{norm } \gamma, \beta\}$$

for $\gamma \in \Gamma^{\bullet}$, $\beta \in F^{\bullet}$. Suppose then that α is equal to the norm of some element γ of $\Gamma^{\bullet}$. Since the element $f(\beta) = \beta$ has an n-th root in Γ, it follows that $\{\gamma, f(\beta)\}$ has an n-th root in $K_2\Gamma$, and therefore that

$$f^*\{\gamma, f(\beta)\} = \{\alpha, \beta\}$$

has an n-th root in K_2F. This completes the proof. ∎

Section 15, which follows, will study the converse problem:

If $\{\alpha,\beta\} \equiv 1 \bmod (K_2F)^n$, *does it follow that* α *is a norm from* $F[x]/(x^n - \beta)$?

We will see that this converse statement is true, for example, when n is square-free and not divisible by the characteristic of F. However the converse statement cannot be true without any restriction on n:

EXAMPLE 14.4. Consider the elements 5 and -1 in the field Q_2 of 2-adic numbers; with $n = 4$. Then

$$\{5, -1\} \equiv 1 \bmod (K_2Q_2)^4,$$

and yet 5 is not a norm from the biquadratic extension field $Q_2(\sqrt[4]{-1})$.

Proof. The congruence $\{5,-1\} \equiv 1 \bmod (K_2Q_2)^4$ can be proved by showing that -1 is a norm from $Q_2(\sqrt[4]{5})$, or can be verified directly by noting that

$$\begin{aligned} \{5,-1\} &= \{5,-1\}\{5,-4\}\{-3,4\} \\ &= \{5,4\}\{-3,4\} = \{-15,4\}, \end{aligned}$$

where the element -15 has a 4-th root in Q_2.

To show that 5 is not a norm, it is only necessary to note that the field $Q_2(\sqrt[4]{-1})$ contains $Q_2(\sqrt{2})$ as subfield. But 5 cannot be a norm from $Q_2(\sqrt{2})$ since the Hilbert symbol $(5,2)_2$ is non-trivial. ∎

§15. Power Norm Residue Symbols

This section will be mainly an exposition of classical material.

Let F be a field containing a primitive n-th root of unity ω. (In other words ω is an element of order precisely n in $F^{\bullet}$.) We will construct a Steinberg symbol a_ω on F with values in the Brauer group of F. (Compare Serre, *Corps locaux,* Hermann 1968.) Our presentation is suggested by O'Meara's treatment of the case $n = 2$ in *Introduction to Quadratic Forms,* Springer 1963.

Here is the basic construction. Given elements α and β of $F^{\bullet}$, let

$$A = A_\omega(\alpha,\beta)$$

be the associative algebra with unit, of dimension n^2 over F, which is generated by two elements x and y subject to the relations

$$x^n = \alpha 1, \quad y^n = \beta 1, \quad yx = \omega xy;$$

where 1 denotes the identity element of A. Thus the monomials $x^i y^j$ with $0 \le i < n$, $0 \le j < n$ are to form a basis for A over F.

The proof that such an algebra exists is not difficult, and will be left to the reader.

Now recall the definition of the Brauer group of F. An algebra A over F (always associative with 1) is *simple* if it has no two-sided ideals other than 0 and A; and *central* if its center is equal to $F1 \cong F$. Wedderburn's theorem asserts that every finite dimensional simple algebra over F is isomorphic to an algebra $M_k(D)$ consisting of all $k \times k$ matrices over some division algebra* $D \supset F$. Furthermore, the integer $k \geq 1$ and the isomorphism class of D are uniquely determined by A. So we can define two such algebras to be *similar* if their associated division algebras are isomorphic over F. The *Brauer group* Br(F) is the abelian group consisting of all similarity classes of finite dimensional central simple algebras over F, using the tensor product over F as composition operation. For further details, see O'Meara, or van der Waerden, *Modern Algebra*, or Weil, *Basic Number Theory*.

THEOREM 15.1. *The algebra* $A = A_\omega(\alpha,\beta)$ *is central simple, and hence represents an element* $a_\omega(\alpha,\beta)$ *of the Brauer group* Br(F). *The resulting function*

$$a_\omega : F^\bullet \times F^\bullet \to \mathrm{Br}(F)$$

is a Steinberg symbol on F.

Proof. Consider the F-linear transformations

$$T_x(a) = xax^{-1}, \quad T_y(a) = yay^{-1}$$

---

* More generally, any simple artinian ring is a matrix algebra over a skew-field. Here is a brief proof. Let $\mathfrak{m} \subset A$ be a minimal right ideal. Since A is simple, the two-sided ideal $A\mathfrak{m}$ must be equal to A; hence the element 1 can be represented as a sum

$$1 \in a_1\mathfrak{m} + a_2\mathfrak{m} + \dots a_k\mathfrak{m}.$$

Choose such a representation with k as small as possible. *Then the* k*-fold direct sum* $\mathfrak{m}\oplus\dots\oplus\mathfrak{m}$ *is isomorphic as right* A*-module to* A *itself.* For the correspondence

$$(m_1,\dots,m_k) \mapsto a_1m_1 + \dots + a_km_k$$

is certainly right A-linear and surjective. If there were a relation of the form $a_1m_1 + \dots + a_km_k = 0$, with say $m_k \neq 0$ and hence $m_kA = \mathfrak{m}$, then the inclusion

$$a_k\mathfrak{m} = a_km_kA \subset a_1\mathfrak{m} + \dots + a_{k-1}\mathfrak{m}$$

would imply that $1 \in a_1\mathfrak{m} + \dots + a_{k-1}\mathfrak{m}$; contradicting the choice of k.

It follows that the ring $\mathrm{End}_A(\mathfrak{m}\oplus\dots\oplus\mathfrak{m})$, consisting of all right A-linear mappings from $\mathfrak{m}\oplus\dots\oplus\mathfrak{m}$ to itself, is isomorphic to the ring $\mathrm{End}_A(A) \cong A$. But $\mathrm{End}_A(\mathfrak{m}\oplus\dots\oplus\mathfrak{m})$ can be identified with the ring of k×k matrices over the skew-field $\mathrm{End}_A(\mathfrak{m})$. This completes the proof. ∎

from A to itself. Each basis vector $x^i y^j$ is an eigenvector both of T_x and T_y, with eigenvalues ω^{-j} and ω^i respectively. It follows immediately that the center of A is spanned over F by the single element $x^0 y^0 = 1$.

To prove that A is simple, consider a non-zero two-sided ideal $\mathfrak{b}$, and choose an element

$$b_0 = \sum \phi_{ij} x^i y^j \in \mathfrak{b}$$

with say $\phi_{pq} \neq 0$. Then the element

$$b_1 = x^{-p} b_0 y^{-q}$$

of $\mathfrak{b}$ can be written as $\sum \Psi_{ij} x^i y^j$ with $\Psi_{00} \neq 0$. Next consider the element

$$b_2 = (T_x - \omega)(T_x - \omega^2) \ldots (T_x - \omega^{n-1}) b_1$$

of $\mathfrak{b}$. Computation shows that b_2 is equal to

$$(1-\omega)(1-\omega^2) \ldots (1-\omega^{n-1}) \sum \Psi_{i0} x^i .$$

Similarly the element

$$b_3 = (T_y - \omega)(T_y - \omega^2) \ldots (T_y - \omega^{n-1}) b_2$$

of $\mathfrak{b}$ is equal to $(1-\omega)^2(1-\omega^2)^2 \ldots (1-\omega^{n-1})^2 \Psi_{00} 1$. Thus b_3 is a unit, and hence $\mathfrak{b} = A$; which proves that A is simple.

To proceed further, we will need the following.

LEMMA 15.2. *Let* A *be a simple algebra of dimension* n^2 *over* F, *and let* $x \in A$ *be an element which satisfies a polynomial equation of the form*

$$f(x) = x^n + \phi_1 x^{n-1} + \ldots + \phi_n 1 = 0 ,$$

but no equation of smaller degree. If f(x) *splits into distinct linear factors over* F, *then* A *is isomorphic to the matrix algebra* $M_n(F)$.

Proof. The subalgebra of A spanned by the powers of x is clearly isomorphic to the quotient ring $F[x]/(f(x))$. By the Chinese Remainder Theorem, this splits as the cartesian product of n copies of F. Hence it contains n mutually annihilating idempotents

$$e_i e_i = e_i, \quad e_i e_j = 0 \text{ for } i \neq j,$$

with $e_1 + \dots + e_n = 1$. Therefore A splits as the direct sum $e_1 A \oplus \dots \oplus e_n A$ of right ideals, at least one of which has dimension $\leq n$ over F. Comparing any proof of Wedderburn's theorem, it follows easily that A is a matrix algebra over F. ∎

EXAMPLE 15.3. Assume once more that F contains a primitive n-th root of unity. If α has an n-th root in F, then the polynomial $x^n - \alpha$ splits into distinct linear factors, so it follows that $a_\omega(\alpha,\beta) = 1$. In particular, this proves that $a_\omega(1,\beta) = 1$.

Proof that the symbol $a_\omega(\alpha,\beta)$ *is bimultiplicative.* The tensor product

$$B = A_\omega(\alpha,\beta) \underset{F}{\otimes} A_\omega(\alpha,\gamma)$$

has generators x, y, X, and Y, subject to the relations

$$x^n = \alpha 1, \quad y^n = \beta 1, \quad yx = \omega xy, \tag{1}$$

$$X^n = \alpha 1, \quad Y^n = \gamma 1, \quad YX = \omega XY, \text{ and} \tag{2}$$

$$xX = Xx, \quad xY = Yx, \quad yX = Xy, \quad yY = Yy. \tag{3}$$

Let B′ be the subalgebra generated by x and yY, and let B″ be the subalgebra generated by $x^{-1}X$ and Y. Then clearly

$$B' \cong A_\omega(\alpha,\beta\gamma), \quad B'' \cong A_\omega(1,\gamma).$$

Note also that each generator of B′ commutes with each generator of B″. Since $B' \otimes B''$ is simple, it follows that the natural map

$$B' \otimes B'' \to B$$

is an isomorphism. This proves that

$$A_\omega(\alpha,\beta\gamma) \otimes A_\omega(1,\gamma) \cong A_\omega(\alpha,\beta) \otimes A_\omega(\alpha,\gamma),$$

or in other words that

$$a_\omega(\alpha,\beta\gamma)\, a_\omega(1,\gamma) = a_\omega(\alpha,\beta)\, a_\omega(\alpha,\gamma).$$

But $a_\omega(1,\gamma) = 1$ by 15.3; so the symbol a_ω is right multiplicative. The left multiplicative law follows immediately, using the identity $a_\omega(\alpha,\beta) = a_{\omega^{-1}}(\beta,\alpha)$.

Proof of the Steinberg identity $a_\omega(1-\beta,\beta) = 1$. Define the "non-commutative binomial coefficient" to be the polynomial

$$b_i^n(c) = f_n(c)/f_i(c)f_{n-i}(c) = c^i b_i^{n-1}(c) + b_{i-1}^{n-1}(c)$$

where

$$f_n(c) = (c-1)(c^2-1)\ldots(c^n-1).$$

It is not difficult to check that this is indeed a polynomial function. Furthermore, if x and y are elements of an arbitrary ring with $yx = xyc$, the element c being in the center, then induction on n shows that

$$(x+y)^n = \sum b_i^n(c)\, x^i y^{n-i}.$$

In particular, for the generators x and y of $A_\omega(\alpha,\beta)$, since $b_i^n(\omega) = 0$ for $0 < i < n$, we obtain

$$(x+y)^n = x^n + y^n = (\alpha+\beta)1.$$

Applying 15.2 to the element $x + y$, it follows that $a_\omega(\alpha,\beta) = 1$ whenever $\alpha + \beta = 1$. This completes the proof of Theorem 15.1. ∎

Recall that $(K_2F)^n$ denotes the subgroup of K_2F consisting of all n-th powers.

COROLLARY 15.4. *The Steinberg symbol* $a_\omega(\alpha,\beta)$ *gives rise to a homomorphism*

$$a_\omega : K_2F \to Br(F)$$

which annihilates the subgroup $(K_2F)^n$.

Proof. As in §11.3, we define this homomorphism by the correspondence

$$\{\alpha,\beta\} \mapsto a_\omega(\alpha,\beta).$$

Since the identity

$$a_\omega(\alpha,\beta)^n = a_\omega(\alpha^n,\beta) = 1$$

follows from 15.3, this completes the proof. ∎

REMARK. It seems possible that the kernel of this homomorphism $a_\omega : K_2F \to Br(F)$ is always precisely equal to $(K_2F)^n$. I do not know how to attack this question.

The manner in which these symbols a_ω depend on the root of unity ω is somewhat strange. It can be described as follows.

LEMMA 15.5. *If* $\xi = \omega^i$ *where* i *is relatively prime to* n *(so that* ξ *is another primitive* n-*th root of unity), then*

$$a_\xi(\alpha,\beta)^i = a_\omega(\alpha,\beta).$$

On the other hand if j *is a divisor of* n, *so that* $\xi = \omega^j$ *is a primitive* (n/j)-*th root of unity, then*

$$a_\xi(\alpha,\beta) = a_\omega(\alpha,\beta)^j.$$

Proof. If i is relatively prime to n, then using the elements x^i and y (in place of x and y) as generators for the algebra $A_\omega(\alpha,\beta)$, we see that

$$A_\omega(\alpha,\beta) \cong A_{\omega^i}(\alpha^i,\beta),$$

from which the first equality follows.

The proof of the second equality can be sketched as follows. A change of generators, similar to that on p. 146, shows that the tensor product $A_\xi(\alpha,\beta) \otimes A_\omega(\beta,\alpha^j)$ is isomorphic to $A_\xi(\alpha,1) \otimes A_\omega(\beta,1)$. Therefore

$$a_\xi(\alpha,\beta)\, a_\omega(\beta,\alpha)^j = a_\xi(\alpha,1)\, a_\omega(\beta,1) = 1,$$

as required. ∎

REMARK 15.6. Fixing the integer n, we can construct a Steinberg symbol which does not depend on any particular choice of ω as follows. Let $Br_n(F)$ denote the subgroup of Br(F) consisting of all elements whose order divides n; and let μ_n be the cyclic group consisting of all n-th roots of unit in F. *Then the symbol*

$$\omega \otimes a_\omega(\alpha,\beta) \in \mu_n \otimes Br_n(F)$$

is independent of the choice of ω. This follows from the computation

$$\omega^i \otimes a_{\omega^i}(\alpha,\beta) = \omega \otimes a_{\omega^i}(\alpha,\beta)^i = \omega \otimes a_\omega(\alpha,\beta),$$

for i prime to n.

Thus there is a well defined associated homomorphism

$$\omega \otimes a_\omega : K_2F \to \mu_n \otimes Br_n(F).$$

Note that the target group $\mu_n \otimes Br_n(F)$ can be identified with $Tor(\mu_n, Br(F))$.

Now we will prove the basic property which suggests the name n-th power *norm residue symbol* for a_ω.

THEOREM 15.7. *The symbol* $a_\omega(\alpha,\beta)$ *is equal to* 1 *if and only if* α *is a norm from the extension field* $F(\sqrt[n]{\beta})$, *or from the extension ring* $F[y]/(y^n-\beta)$.

The two statements are equivalent since, if the polynomial $y^n-\beta$ splits as the product of d irreducible factors, then the ring $F[y]/(y^n-\beta)$ splits as the cartesian product of d fields, each of which is clearly isomorphic to $F(\sqrt[n]{\beta})$.

*Proof.** If $a_\omega(\alpha,\beta) = 1$, then $A_\omega(\alpha,\beta) \cong M_n(F)$ is isomorphic to the ring $Hom_F(V,V)$ where V is an n-dimensional vector space. Thus the generators x and y of $A_\omega(\alpha,\beta)$ correspond to linear transformations X and Y of V. The minimal polynomial $y^n-\beta$ of Y has degree n, hence we can choose a basis $v_1,\ldots,v_n$ for V so as to put Y in "companion matrix" normal form. In other words

$$Y(v_i) = v_{i+1} \text{ for } i < n, \quad Y(v_n) = \beta v_1 .$$

Consider the F-linear transformation $Z \mapsto T_Y(Z) = YZY^{-1}$ of $Hom_F(V,V)$. The element Z defined by

$$Z(v_i) = \omega^i v_i$$

is clearly an eigenvector of T_Y with eigenvalue ω^{-1}. Since the ω^{-1}-eigenspace is spanned by the elements $X^{-1}, X^{-1}Y, \ldots, X^{-1}Y^{n-1}$, it follows that we can write

$$Z = X^{-1}f(Y)$$

for some polynomial f. Hence $Z^n = I$ is equal to

$$X^{-1}f(Y)X^{-1}f(Y) \ldots X^{-1}f(Y) = f(\omega Y)f(\omega^2 Y) \ldots f(\omega^n Y)X^{-n} .$$

Therefore the n-fold product $\prod f(\omega^i Y)$ is equal to $X^n = \alpha I$.

---

* Compare van der Waerden, *Modern Algebra* II, p. 215.

Now consider an extension field $F(\eta)$ where $\eta^n = \beta$. Mapping Y to η, this proves that

$$f(\omega\eta)\, f(\omega^2\eta) \ldots f(\omega^n\eta) = \alpha .$$

If $F(\eta)$ has degree n over F, then clearly this product is just the norm of $f(\eta)$. More generally, if $F(\eta)$ has degree d over F, then it is easily seen that α is equal to the norm of the (n/d)-fold product

$$f(\omega\eta)\, f(\omega^2\eta) \ldots f(\omega^{n/d}\eta) .$$

Thus α is a norm from $F(\eta)$.

The converse statement can be proved directly, or can be derived from §14 as follows. If α is a norm from $F[y]/(y^n-\beta)$, then

$$\{\alpha,\beta\} \equiv 1 \mod (K_2F)^n$$

by 14.3; hence $a_\omega(\alpha,\beta) = 1$ by 15.4. This completes the proof. ∎

The Case of a Local Field

To illustrate these constructions, suppose that F is a *local field*; that is suppose that F is complete under a discrete valuation v, with finite residue class field.

LEMMA 15.8. *If* F *is a local field containing a primitive* n*-th root of unity* ω, *then there exist field elements* π *and* γ *so that the element* $a_\omega(\pi,\gamma)$ *in the Brauer group has order precisely* n.

Proof. Let $E \supset F$ be the unique unramified extension of degree n. Since this is a Galois extension with Galois group cyclic of order n, and since F contains the n-th roots of unity, it follows from Kummer theory that $E = F(\sqrt[n]{\gamma})$ for some element γ in $F^\bullet$.

(As a reference for Kummer theory, see Lang, *Algebra*, Addison-Wesley 1965. For the properties of local fields used here, see Artin, *Algebraic Numbers and Algebraic Functions*, Gordon and Breach 1967, pp. 127-130.)

Since this extension is unramified, the image of the norm homomorphism $E^\bullet \to F^\bullet$ must consist entirely of elements α satisfying

$$v(\alpha) \equiv 0 \pmod{n} .$$

Hence, if π is any element of $F^{\bullet}$ with $v(\pi) = 1$, the powers $\pi, \pi^2, \ldots, \pi^{n-1}$ cannot be norms from $F(\sqrt[n]{\gamma})$. Using 15.7, it follows that the element

$$a_\omega(\pi,\gamma) \in \mathrm{Br}(F)$$

has order precisely equal to n. This completes the proof. ∎

REMARK 15.9. The customary n-th power norm residue symbol on a local field F takes values in the group μ_n of n-th roots of unity in F. This can be related to our definitions as follows. The Brauer group Br(F) of a local field is canonically isomorphic to the group Q/Z of rational numbers modulo 1. (See for example Serre's exposition in Cassels and Fröhlich, *Algebraic Number Theory*, p. 130.) Hence the subgroup $\mathrm{Br}_n(F)$ of 15.6 is cyclic of order n, with a preferred generator b_0. The required symbol $(\alpha,\beta)_F$ with values in μ_n can now be defined as follows. *Let* $(\alpha,\beta)_F$ *be the unique element of* μ_n *such that the equation*

$$\omega \otimes a_\omega(\alpha,\beta) = (\alpha,\beta)_F \otimes b_0$$

is valid in the group $\mu_n \otimes \mathrm{Br}_n(F)$. It follows from 15.6 that this Steinberg symbol $(\alpha,\beta)_F$ is well defined.

[If we assume the result that $\mathrm{Br}_n(F)$ is cyclic, then the preferred generator b_0 can be constructed as follows. Let $E = F(\sqrt[n]{\gamma})$ be the unique unramified extension of degree n, as in 15.8. The *Frobenius automorphism* ϕ of E over F is defined by the requirement that ϕ induces the automorphism $\bar{e} \mapsto \bar{e}^q$ on the residue class field $\bar{E}$, where q denotes the number of elements in $\bar{F}$. Setting

$$\omega(\gamma) = \phi(\sqrt[n]{\gamma})/\sqrt[n]{\gamma},$$

we obtain a primitive n-th root of unity. Now let

$$b_0 = a_{\omega(\gamma)}(\pi,\gamma).$$

This symbol generates $\mathrm{Br}_n(F)$ by 15.8, and is independent of the choice of γ by 15.5. To prove that it does not depend on the choice of π, one must first show that every $u \in F$ with $v(u) = 0$ is a norm from E. This can be verified by an elementary argument, using the fact that the norm and trace from $\bar{E}$ to $\bar{F}$ are both surjective. The required equation $a_{\omega(\gamma)}(\pi u,\gamma) = a_{\omega(\gamma)}(\pi,\gamma)$ then follows from 15.7.]

REMARK 15.10. Here is an outline of the more usual definition of the n-th power norm residue symbol in a local field F. Let $K \supset F$ be the Kummer extension obtained by adjoining the n-th roots of all elements of $F^\bullet$. This is a finite extension since $(F^\bullet)^n$ is a subgroup of finite index in $F^\bullet$. By Kummer theory, the Galois group G of K over F is canonically isomorphic to $\mathrm{Hom}(F^\bullet, \mu_n)$. (Each automorphism σ in G corresponds to the homomorphism $\beta \mapsto \sigma(\sqrt[n]{\beta})/\sqrt[n]{\beta}$ from $F^\bullet$ to μ_n.) Combining this Kummer isomorphism with the norm residue isomorphism

$$F^\bullet/\mathrm{norm}\, K^\bullet \xrightarrow{\cong} G$$

of local class field theory, we obtain a composite homomorphism

$$F^\bullet \to F^\bullet/\mathrm{norm}\, K^\bullet \cong G \cong \mathrm{Hom}(F^\bullet, \mu_n)$$

which is adjoint to the required bimultiplicative pairing

$$F^\bullet \times F^\bullet \to \mu_n .$$

The Steinberg identity $(1-\beta, \beta) \mapsto 1$ is proved by noting that $1-\beta$ is a norm from the field $E = F(\sqrt[n]{\beta})$, and then using the commutativity of the diagram

$$\begin{array}{ccccc} F^\bullet & \to & F^\bullet/\mathrm{norm}_{K/F} K^\bullet & \xrightarrow{\cong} & G_{K/F} \\ & & \downarrow & & \downarrow \\ & & F^\bullet/\mathrm{norm}_{E/F} E^\bullet & \xrightarrow{\cong} & G_{E/F} . \end{array}$$

For further details, the reader is referred to the following:

E. Artin and J. Tate, *Class Field Theory*, Benjamin 1968;

J. W. S. Cassels and A. Fröhlich (editors), *Algebraic Number Theory*, Thompson Book Co. and Academic Press 1967;

J.-P. Serre, *Corps locaux*, Hermann 1968.

A. Weil, *Basic Number Theory*, Springer 1967.

The Quotient $K_2F/(K_2F)^n$

Now let us return to the question studied in §§14.3 and 14.4.

COROLLARY 15.11. *If* F *contains the* n-*th roots of unity, then*

$$\{\alpha, \beta\} \equiv 1 \ \mathrm{mod}(K_2F)^n$$

if and only if α *is a norm from* $F[y]/(y^n - \beta)$.

Proof. If $\{\alpha,\beta\} \equiv 1 \mod(K_2F)^n$, then $a_\omega(\alpha,\beta) = 1$ by 15.4, hence α is a norm by 15.7. Since the converse statement was proved in § 14.3, this completes the proof. ∎

THEOREM 15.12 (Bass, Tate). *Let* n *be any integer which is square-free, and not divisible by the characteristic of* F. *Then*

$$\{\alpha,\beta\} \equiv 1 \mod(K_2F)^n$$

if and only if α *is a norm from* $F[y]/(y^n-\beta)$.

(Compare §14.4.)

Proof. Let p be a prime divisor of n. If $\{\alpha,\beta\} \equiv 1 \mod(K_2F)^n$, then certainly $\{\alpha,\beta\} \equiv 1 \mod(K_2F)^p$. We will show that α is a norm from $F(\eta')$ where η' is a suitably chosen p-th root of β. Let η be an arbitrary p-th root of β. (These elements η, η' are to belong to some fixed splitting field of the polynomial $y^n-\beta$.)

Case 1. If $F(\eta)$ has degree $d < p$ over F, then we will show that β has a p-th root η' in F itself. In fact, setting $ip + jd = 1$, let

$$\eta' = \beta^i \operatorname{norm}_{F(\eta)/F} \eta^j .$$

The computation

$$\beta^{ip} \operatorname{norm} \eta^{jp} = \beta^{ip} \operatorname{norm} \beta^j = \beta^{ip+jd} = \beta$$

shows that η' is indeed a p-th root of β. Certainly α is a norm from the field $F(\eta') = F$.

Case 2. If $F(\eta)$ has degree p over F, then we will make use of the field E which is obtained by adjoining the p-th roots of unity to F. Let $d < p$ be the degree of E over F.

Since $\{\alpha,\beta\}$ has a p-th root in K_2F, its image in K_2E has a p-th root also. Using 15.11, it follows that α is a norm from $E(\eta)$ to E. Hence, using the commutative diagram

$$\begin{array}{ccc} E(\eta)^\bullet & \longrightarrow & E^\bullet \\ \downarrow & & \downarrow \\ F(\eta)^\bullet & \longrightarrow & F^\bullet \end{array}$$

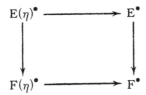

of norm homomorphisms, we see that $\mathrm{norm}_{E/F}\ \alpha = \alpha^d$ is the norm of some element $\xi \in F(\eta)^{\bullet}$. Setting $ip + jd = 1$, it follows that α is the norm of $\alpha^i \xi^j \in F(\eta)^{\bullet}$.

Thus, for each of the primes p_k dividing n, we can find a p_k-th root of β, call it η_k, so that α is a norm from $F(\eta_k)$, and so that $F(\eta_k)$ has degree either 1 or p_k over F. Since these degrees are relatively prime, it follows easily that α is a norm from the compositum $F(\eta_1, \ldots, \eta_r) = F(\eta_1 \eta_2 \ldots \eta_r)$. But this compositum is a direct factor of the ring $F[y]/(y^n - \beta)$. Therefore α is a norm from $F[y]/(y^n - \beta)$.

Since the converse statement was proved in §14, this completes the proof of 15.12. ■

§16. Number Fields

Let F be a finite extension of the field of rational numbers, and let m be the number of roots of unity in F. Thus the group $\mu(F)$, consisting of all roots of unity in F, is cyclic of order m.

Calvin Moore has proved a uniqueness theorem for the m-th power reciprocity formula in F. This section will state the reciprocity formula and Moore's uniqueness theorem without proof. (See 16.1.) As a corollary to Moore's theorem, it is shown that the group $SL(\Lambda)$, where Λ denotes the ring of integers of F, is generated by elementary matrices. (A more direct proof of this fact is also sketched.) Therefore the exact sequence of §13 takes the form

$$K_2F \to \bigoplus K_1\Lambda/\mathfrak{p} \overset{0}{\to} K_1\Lambda \to K_1F \to \dots .$$

REMARK. Further information about this sequence has been obtained by Bass and Tate, and more recently by H. Garland. If n is sufficiently large, Garland shows that the Schur multiplier $H_2SL(n,\Lambda)$ is finite. Taking the direct limit as $n \to \infty$, it follows that $K_2\Lambda$ is a torsion group.* Further, in the exact sequence $K_2\Lambda \to K_2F \to \bigoplus K_1\Lambda/\mathfrak{p} \overset{0}{\to} \dots$, he shows that the image of $K_2\Lambda$ is finite. In the totally real case, Birch and Tate conjecture that the order of this finite image is equal to the value of the zeta function of F at -1 multiplied by the number of roots of unity in the extension field $F(\sqrt{F})$ obtained by adjoining the square roots of all elements of F. Compare Tate's paper in the Proceedings of the International Congress at Nice.

---

* K. Dennis has recently shown that $H_2SL(n,\Lambda)$ maps onto $H_2SL(n+1,\Lambda)$ for $n \geq 5$. Therefore $K_2\Lambda$ is actually finite.

To begin our discussion of Moore's theorem, let v denote either a discrete valuation of the field F (as in §11), or an archimedean absolute value extending the classical absolute value $|q| = \mathrm{Max}(q, -q)$ on the rational numbers. Then the completion F_v is either a local field, or the field of real numbers, or the field of complex numbers. In the local and real cases the group $\mu(F_v)$, consisting of all roots of unity in the completed field, is finite cyclic. But in the complex case the group $\mu(F_v) \cong \mu(C)$ is certainly not a finite group. We will systematically exclude the complex case henceforth, and assume that F_v is either a local field or the real field.

Let $m(v)$ denote the order of the finite cyclic group $\mu(F_v)$. The $m(v)$-th *power norm residue symbol* $(x,y)_v$ is a continuous Steinberg symbol on F_v with values in $\mu(F_v)$. For the definition of this symbol, see §15.9 or 15.10 and the references listed there. It is important that we exclude the complex archimedean case: In fact Moore has shown that any continuous Steinberg symbol on the complex field is identically equal to 1. (See the Appendix.)

Now suppose that x and y are non-zero elements of the field $F \subset F_v$. In order to put all of the various symbols $(x,y)_v$ into one group, we apply the surjection

$$\mu(F_v) \to \mu(F)$$

which carries each root of unity ξ to the power $\xi^{m(v)/m}$. The m-th power *reciprocity theorem* then asserts that

$$\prod_v (x,y)_v{}^{m(v)/m} = 1 .$$

(See for example Artin and Tate, *Class Field Theory*, p. 171 or Cassels and Fröhlich, *Algebraic Number Theory*, pp. 269, 352.) Here the product is to be taken over all discrete valuations of F, and all archimedean absolute values with $F_v \cong R$.

Moore's uniqueness theorem can be stated as follows. Let S^1 be the circle group, consisting of all complex numbers of absolute value 1.

Suppose that the $m(v)$-th *power residue symbols* $(x,y)_v$ *satisfy a relation*

$$\prod_v h_v((x,y)_v) = 1$$

for every x *and* y *in* F, *the* h_v *being homomorphisms from* $\mu(F_v)$ *to the circle group* S^1. *Then there exists a homomorphism* h *from* $\mu(F)$ *to* S^1 *so that*

$$h_v(\xi) = h(\xi^{m(v)/m})$$

for every v.

For the proof, the reader is referred to Moore, *Group extensions of p-adic and adelic linear groups,* Publ. Math. I.H.E.S. 35, Theorem 7.4.

Conversely of course, given any $h : \mu(F) \to S^1$, if we set $h_v(\xi) = h(\xi^{m(v)/m})$, then it is clear that the product of the $h_v((x,y)_v)$ must equal 1. Hence, making use of §11.3, Moore's statement can be expressed succinctly by the exact sequence

(1) $$\mathrm{Hom}(\mu(F),S^1) \to \mathrm{Hom}(\bigoplus \mu(F_v),S^1) \to \mathrm{Hom}(K_2F,S^1).$$

Another completely equivalent formulation is the following.

RECIPROCITY UNIQUENESS THEOREM 16.1. *The sequence*

$$K_2F \to \bigoplus_v \mu(F_v) \to \mu(F) \to 1$$

is exact; where the first homomorphism carries each generator $\{x,y\}$ *of* K_2F *to the element whose* v-*th component is the* m(v)-*th power residue symbol* $(x,y)_v$, *and the second homomorphism carries each element* $\{\xi_v\}$ *to the product* $\prod_v \xi_v^{m(v)/m}$. *Here* v *ranges over all discrete valuations, and all archimedean absolute values with completion* $F_v \cong R$.

Proof. Since the target group S^1 is injective, and contains elements of all orders, the exactness of this sequence is completely equivalent to the exactness of the dual sequence (1). This completes the proof. ∎

Here is an application of 16.1.

COROLLARY 16.2. *Let* Λ *be any Dedekind domain whose quotient field* F *is a finite extension of the rationals. Then the homomorphism*

$$d : K_2F \to \bigoplus K_1\Lambda/\mathfrak{p}$$

of §13.8 *is surjective.*

For example Λ could be the ring of algebraic integers of F.

Proof. Note that every non-zero prime ideal $\mathfrak{p} \subset \Lambda$ gives rise to a unique $\mathfrak{p}$-adic valuation of F. Let $F_{\mathfrak{p}}$ denote the $\mathfrak{p}$-adic completion, and let $m(\mathfrak{p})$ be the order of $\mu(F_{\mathfrak{p}})$.

The tame symbol of §11.5, which is used to define the homomorphism $K_2F \to K_1\Lambda/\mathfrak{p}$, can be expressed as the image of the $m(\mathfrak{p})$-th power norm residue symbol $(x,y)_{\mathfrak{p}}$ under an appropriate surjection

$$\psi_{\mathfrak{p}} : \mu(F_{\mathfrak{p}}) \to K_1\Lambda/\mathfrak{p}\,.$$

(Compare Serre, *Corps Locaux*, p. 217, or Artin and Tate, p. 170. In fact, we will see in the appendix that any continuous Steinberg symbol on a local field is the image of the norm residue symbol under a suitable homomorphism.) The proof of 16.2 now proceeds as follows.

Case 1. Suppose that no rational prime is a unit in Λ. (E.g., suppose that Λ is precisely the ring of integers in F.) Let us start with an arbitrary element $\{x_{\mathfrak{p}}\}$ of the direct sum $\bigoplus K_1\Lambda/\mathfrak{p}$. Choosing roots of unity $\xi_{\mathfrak{p}}$ in $\mu(F_{\mathfrak{p}})$, almost all equal to 1, with $\psi_{\mathfrak{p}}(\xi_{\mathfrak{p}}) = x_{\mathfrak{p}}$, we obtain a corresponding element $\{\xi_{\mathfrak{p}}\}$ of the direct sum $\bigoplus \mu(F_{\mathfrak{p}})$. Consider the product

$$\xi = \prod \xi_{\mathfrak{p}}{}^{m(\mathfrak{p})/m}$$

in $\mu(F)$. By altering the choices of $\xi_{\mathfrak{p}}$, if necessary, we will try to set ξ equal to 1.

Let q be any rational prime dividing $m = |\mu(F)|$, and let $\mathfrak{q}$ be any prime ideal of Λ which contains q. Then the cyclic group $K_1\Lambda/\mathfrak{q}$ has order prime to q, so the surjection

$$\psi_{\mathfrak{q}} : \mu(F_{\mathfrak{q}}) \to K_1\Lambda/\mathfrak{q}$$

certainly annihilates the q-primary component of the cyclic group $\mu(F_{\mathfrak{q}})$. But $\mu(F_{\mathfrak{q}})$ maps onto $\mu(F)$. So, by altering the q-primary component of $\xi_{\mathfrak{q}}$, we can kill the q-primary component of $\xi = \prod \xi_{\mathfrak{p}}{}^{m(\mathfrak{p})/m}$ without changing the other primary components.

Repeating this procedure for each prime divisor of m, it follows that we can lift any element $\{x_\mathfrak{p}\}$ of $\bigoplus K_1\Lambda/\mathfrak{p}$ to an element $\{\xi_\mathfrak{p}\}$ of $\bigoplus \mu(F_\mathfrak{p})$ which satisfies the condition

$$\prod_\mathfrak{p} \xi_\mathfrak{p}^{m(\mathfrak{p})/m} = 1 .$$

Hence, using 16.1, we can lift $\{x_\mathfrak{p}\}$ to an element of the group K_2F. This completes the proof in Case 1.

Case 2. Suppose that some discrete valuation v_0 of F is not a $\mathfrak{p}$-adic valuation for any prime ideal $\mathfrak{p} \subset \Lambda$. Choosing $\{\xi_\mathfrak{p}\}$ as above, we can then choose an element ξ_{v_0} of $\mu(F_{v_0})$ so that the product

$$\xi_{v_0}^{\ m(v_0)/m} \prod \xi_\mathfrak{p}^{\ m(\mathfrak{p})/m}$$

is equal to 1. The proof then proceeds as in Case 1.

But Cases 1 and 2 clearly exhaust all of the possibilities. For if some rational prime p is a unit in Λ, then the p-adic valuation of the rational numbers extends to the required valuation v_0 of F. This completes the proof of 16.2. ∎

Now consider the sequence

$$K_2F \to \bigoplus K_1\Lambda/\mathfrak{p} \to K_1\Lambda \to K_1F \to \dots$$

of §13. If F is a number field, we have shown that the first homomorphism is surjective. Since the composition of the first two is zero by 13.8, it follows that *the second homomorphism in this sequence is zero.* But the image of the second homomorphism is the subgroup

$$SK_1\Lambda = SL(\Lambda)/E(\Lambda).$$

This proves:

COROLLARY 16.3. *If* Λ *is a Dedekind domain, with quotient field finite over the rationals, then* $SK_1\Lambda = 0$. *In other words the infinite special linear group* $SL(\Lambda)$ *is generated by elementary matrices.*

REMARK. It follows that $SL(n,\Lambda)$ is generated by elementary matrices for every $n \geq 3$. For it is proved in Bass, Milnor and Serre, Publ. Math. I.H.E.S. 33 p. 120, or in Bass, *Algebraic K-Theory*, pp. 238-240, or in Vaserštein, Math. of U.S.S.R. Sbornik 8 p. 387, that for any Dedekind domain Λ and any $n \geq 3$ the subgroup

$$E(n,\Lambda) \subset SL(n,\Lambda)$$

is normal, with quotient group isomorphic to $SK_1\Lambda$. (The corresponding statement for $n = 2$ is false. For example if $\Lambda = Z[\sqrt{-5}]$ then Swan has shown that the subgroup $E(2,\Lambda) \subset SL(2,\Lambda)$ is abnormal, and that its normal closure has infinite index in $SL(2,\Lambda)$. See Bull. Amer. Math. Soc. 74 (1968), p. 580; or Advances in Math. 6 (1971), 1-77.)

If Λ is precisely the ring of algebraic integers in F, then a direct proof of 16.3 can be given as follows. (Compare Bass, Milnor and Serre, p. 77.)

Alternative proof. We will need the classical theorem of Hurwitz which asserts that $SL(2,\Lambda)$ is finitely generated, Λ being the ring of integers in the number field F. (See Hurwitz, *Math. Werke* 2, 244-268.) Since $SL(2,\Lambda)$ maps onto $SL(\Lambda)/E(\Lambda) = SK_1\Lambda$ by §13.6, it follows that the abelian group $SK_1\Lambda$ is finitely generated. In fact $SK_1\Lambda$ is finite. For given any generator $\begin{bmatrix} b \\ a \end{bmatrix}$ we can choose an integer $r > 0$ so that $b^r \equiv 1 \bmod a$, and conclude that $\begin{bmatrix} b \\ a \end{bmatrix}^r = 1$, using §13.2.

Suppose first that Λ contains the p-th roots of unity for some prime p. We will show that the finite group $SK_1\Lambda$ has no p-torsion. Clearly it suffices to show that every Mennicke symbol $\begin{bmatrix} b \\ a \end{bmatrix}$ has a p-th root in $SK_1\Lambda$.

Using the Chinese remainder theorem, the element a is congruent modulo Λb to some a_1 which is relatively prime to p. In fact it suffices to choose a_1 so that

$a_1 \equiv a \quad \bmod b$, and

$a_1 \equiv 1 \quad \bmod \mathfrak{p}$

for every prime $\mathfrak{p}$ which divides p but not b. Then $\begin{bmatrix} b \\ a \end{bmatrix} = \begin{bmatrix} b \\ a_1 \end{bmatrix}$.

We will make use of the p-th power norm residue symbols $(x,y)_v \in \mu_p$ and the reciprocity formula $\prod_v (x,y)_v = 1$. Choosing some fixed prime $\mathfrak{q}$ dividing p, note the following.

ASSERTION. *There exist two units* u_0 *and* w_0 *in the* $\mathfrak{q}$*-adic completion of* Λ *so that the* p*-th power norm residue symbol* $(u_0,w_0)_{\mathfrak{q}}$ *is non-trivial.*

Proof. Let $U = \Lambda_{\mathfrak{q}}^\bullet$ be the group of units in the completion. Since U contains the p-th roots of unity, and since p is equal to the residue class field characteristic, it follows that the subgroup $U^p \subset U$ has index at least p^2. (See A.5 in the Appendix; or Lang, *Algebraic Number Theory,* p. 47.) Let π be a prime element of $\Lambda_{\mathfrak{q}}$ (that is, let $v_{\mathfrak{q}}(\pi) = 1$). The subgroup $U_0 \subset U$ consisting of all elements u for which $(u,\pi)_{\mathfrak{q}} = 1$ has index at most p, so there exists some u_0 in U_0 which has no p-th root in $F_{\mathfrak{q}}^\bullet$. But this implies that

$$(u_0,y)_{\mathfrak{q}} \neq 1$$

for some element y in $F_{\mathfrak{q}}^\bullet$. (Compare Serre, *Corps Locaux,* p. 215. This statement follows easily either from §15.10 or from §A.13.) Setting $y = \pi^i w_0$, it follows that $(u_0,w_0)_{\mathfrak{q}} \neq 1$.

The alternative proof of 16.3 now proceeds as follows. By the Chinese remainder theorem there exists an element b_2 satisfying:

$b_2 \equiv b \mod a_1$.

$b_2 \equiv w_0 \mod \mathfrak{q}^N$, and

$b_2 \equiv 1 \mod \mathfrak{p}^N$ for every $\mathfrak{p}|p$ with $\mathfrak{p} \neq \mathfrak{q}$.

Here the integer N must be large enough so that b_2/w_0 has a p-th root in the completion $F_{\mathfrak{q}}$, and so that b_2 has a p-th root in $F_{\mathfrak{p}}$ for each $\mathfrak{p} \neq \mathfrak{q}$ dividing p. (Compare §A.4.)

Now we invoke the generalized Dirichlet theorem, which asserts the following. *The arithmetic progression consisting of all* b_2 *satisfying the above congruences contains an element which is "prime" in the sense that the ideal* Λb_2 *which it spans is a prime ideal.* (See for example Hasse's

Zahlbericht, Jahresber. Deut. Math. Ver. 35 (1926), §8.) *Further this prime element* b_2 *can be chosen so as to be positive in every real completion of* F.

Now note that the p-th power norm residue symbol $(u_0,b_2)_{\mathfrak{q}}$ is equal to $(u_0,w_0)_{\mathfrak{q}} \neq 1$. Hence there exists some power $u = (u_0)^i$ so that the equality

$$(a_1,b_2)_{\Lambda b_2}(u,b_2)_{\mathfrak{q}} = 1$$

is satisfied. Choose a "prime element" a_3 so that

$$a_3 \equiv a_1 \mod b_2, \text{ and}$$

$$a_3 \equiv u \mod \mathfrak{q}^N$$

with N as above. Then

$$\begin{bmatrix} b_2 \\ a_3 \end{bmatrix} = \begin{bmatrix} b_2 \\ a_1 \end{bmatrix} = \begin{bmatrix} b \\ a_1 \end{bmatrix} = \begin{bmatrix} b \\ a \end{bmatrix}.$$

Let us look at the reciprocity formula

$$\prod_v (a_3,b_2)_v = 1 .$$

The symbols $(a_3,b_2)_v$ corresponding to real completions, or corresponding to primes $\mathfrak{p} \neq \mathfrak{q}$ which divide p, are trivial by the choice of b_2. For any prime $\mathfrak{r}$ not dividing p, the symbol $(a_3,b_2)_{\mathfrak{r}}$ is the homomorphic image of the tame symbol of §11.5 (compare §A.8, or *Corps Locaux,* p. 217), and hence is trivial unless $\mathfrak{r}$ divides a_3 or b_2. Thus the reciprocity formula reduces to

$$(a_3,b_2)_{\Lambda a_3}(a_3,b_2)_{\Lambda b_2}(a_3,b_2)_{\mathfrak{q}} = 1 .$$

But the last two factors of this product are equal to $(a_1,b_2)_{\Lambda b_2}$ and $(u,b_2)_{\mathfrak{q}}$ respectively, with product 1. Therefore

$$(a_3,b_2)_{\Lambda a_3} = 1 ,$$

which means that b_2 is congruent to a p-th power modulo a_3. Thus

$$b_2 \equiv x^p \mod a_3$$

and hence

$$\begin{bmatrix} b \\ \\ a \end{bmatrix} = \begin{bmatrix} b_2 \\ \\ a_3 \end{bmatrix} = \begin{bmatrix} x \\ \\ a_3 \end{bmatrix}^p .$$

Thus if F contains the p-th roots of unity, then every Mennicke symbol has a p-th root, hence $SK_1\Lambda$ has no p-torsion. Now suppose that F does not contain the p-th roots of unity. Let Γ be the ring of integers in the extension field $F(e^{2\pi i/p})$. Consider the inclusion homomorphism $\iota_* : K_1\Lambda \to K_1\Gamma$ and the transfer $\iota^* : K_1\Gamma \to K_1\Lambda$, and recall that the composition

$$\iota^* \circ \iota_* : K_1\Lambda \to K_1\Lambda$$

is given by multiplication with the element

$$\iota^*[\Gamma] \in K_0\Lambda .$$

Clearly $\iota^*[\Gamma] = d + \gamma$ where d denotes the degree of $F(e^{2\pi i/p})$ over F, and $\gamma \in \tilde{K}_0\Lambda$ denotes the projective class of Γ over Λ.

Thus if we start with an element α of order p in $SK_1\Lambda$, then $\iota_*(\alpha)$ has order p in $SK_1\Gamma$, hence $\iota_*(\alpha) = 0$. Therefore $(d+\gamma)\alpha = 0$, and hence

$$d^2\alpha = (d-\gamma)\,(d+\gamma)\alpha = 0 ,$$

using §1.11. But d is relatively prime to p, so this proves that $\alpha = 0$.

Thus the finite group $SK_1\Lambda$ has no p-torsion for any prime p, hence $SK_1\Lambda = 0$. This completes the alternative proof. ■

Appendix

Continuous Steinberg Symbols

Calvin Moore has given a complete description of continuous Steinberg symbols on the classical locally compact fields. If F is either a local field or the field of real numbers, then every continuous Steinberg symbol on F is a homomorphic image of the m-th power norm residue symbol, m being the number of roots of unity in F. On the other hand, for the field C of complex numbers, every continuous Steinberg symbol on C is trivial. This appendix contains proofs of these statements. The following hypothesis will suffice for our arguments.

DEFINITION. A Steinberg symbol c on a topological field F will be called *weakly continuous* if for each $\beta \in F^\bullet$ the annihilator, consisting of all α with $c(\alpha,\beta) = 1$, forms a closed subset of $F^\bullet$.

This definition makes no mention of a topology for the target group A of c. However if A is a Hausdorff space and the function

$$c : F^\bullet \times F^\bullet \to A$$

is continuous, then clearly c is weakly continuous.

The Real and Complex Fields

First consider a Steinberg symbol $c(\alpha,\beta)$ on the field R of real numbers. The identities

$$c(3,-2) = c(3,-3) = 1$$

show that $c(3,\beta) = 1$ for every rational number β of the form $(-2)^i(-3)^j$. But the set of all such rationals clearly forms a dense subgroup of $R^\bullet$. So if the Steinberg symbol c is weakly continuous, then it follows that $c(3,\beta) = 1$ for all $\beta \in R^\bullet$. Similarly the identities

$$c(4,-3) = c(4,-4) = 1$$

imply that $c(4,\beta) = 1$ for all β. But every positive real number can be approximated arbitrarily closely by a product $3^i 4^j$, so it follows that $c(\alpha,\beta) = 1$ whenever $\alpha > 0$. Evidently this proves the following.

THEOREM A.1. *If* $c(\alpha,\beta)$ *is a weakly continuous Steinberg symbol on the field* R *of real numbers, then* $c(\alpha,\beta) = 1$ *if* α *or* β *is positive, and*

$$c(\alpha,\beta) = c(-1,-1)$$

has order at most 2 *if* α *and* β *are both negative.*

In other words, $c(u,v)$ is the image of the quadratic norm residue symbol $(u,v)_R$ under the homomorphism $\{\pm 1\} \to A$ which carries $(-1,-1)_R$ to $c(-1,-1)$.

COROLLARY A.2. *Every weakly continuous Steinberg symbol on the complex field is trivial.*

For if ω is a root of unity, say $\omega^n = 1$, then for any β in $C^\bullet$ we can choose an n-th root, $\beta = \gamma^n$, and conclude that

$$c(\omega,\beta) = c(\omega,\gamma^n) = c(\omega^n,\gamma) = 1 .$$

It follows by weak continuity that $c(\omega',\beta) = 1$ for every number ω' on the unit circle. Hence, for any α,

$$c(\alpha,\beta) = c(|\alpha|, \beta) = c(|\alpha|, |\beta|) = 1$$

using A.1, which completes the proof. ∎

Local Fields

Assume now that F is complete under a discrete valuation v with finite residue class field $\bar{F}$. Let Λ be the valuation ring of F, and $\mathfrak{P} = \Lambda - \Lambda^\bullet$ the prime ideal. The number of elements in $\bar{F} = \Lambda/\mathfrak{P}$ will be denoted by $q = p^f$, where p is prime. If F has characteristic zero, let $e = v(p)$ be the ramification index of F over the field of p-adic numbers. In the characteristic p case, we set $e = v(0) = +\infty$.

We begin with some well known properties of the multiplicative group $F^\bullet$. For each $i \geq 1$ let

$$U_i = 1 + \mathfrak{P}^i$$

denote the subgroup of $F^\bullet$ consisting of all elements congruent to 1 modulo $\mathfrak{P}^i$. Note that the quotient group U_i/U_{i+1} is isomorphic to the additive group of the residue class field. Let $\mu_{q-1} \subset \Lambda^\bullet$ be the cyclic group consisting of all $(q-1)$-st roots of unity. Let π denote some generator for the prime ideal $\mathfrak{P}$. (In other words let $v(\pi) = 1$.)

LEMMA A.3. *Every element of* $F^\bullet$ *can be written uniquely as a product* $\pi^i \eta u_1$ *where* η *belongs to the cyclic group* μ_{q-1} *of order* $q-1$ *and* u_1 *belongs to* U_1. *If* n *is relatively prime to* p, *then every element of* U_i *has a unique* n*-th root in* U_i.

Proof. The last statement follows from the fact that U_i is a "pro-p-group." That is, it is the inverse limit of a sequence

$$U_i/U_{i+1} \leftarrow U_i/U_{i+2} \leftarrow U_i/U_{i+3} \leftarrow \cdots$$

of finite p-groups. In particular this argument shows that every element of U_1 has a unique $(q-1)$-st root. Hence the exact sequence

$$1 \to U_1 \to \Lambda^\bullet \to (\Lambda/\mathfrak{P})^\bullet \to 1$$

splits uniquely. The proof is now straightforward.

The existence of p-th roots is more difficult. It will be convenient to set $e_0 = e/(p-1)$. We will say that F *contains the* p*-th roots of unity* if it contains p distinct p-th roots of 1.

LEMMA A.4. *The correspondence* $u \mapsto u^p$ *induces an isomorphism* $U_i \to U_{i+e}$ *for* $i > e_0$. *If* i *is precisely equal to* e_0, *then the* p*-th power homomorphism*

$$U_{e_0} \to U_{e_0+e} = U_{pe_0}$$

either has kernel and cokernel of order p, *or is an isomorphism, according as* F *does or does not contain the* p*-th roots of unity. Finally, if* $i < e_0$, *then the correspondence* $u \mapsto u^p$ *induces an isomorphism*

$$U_i/U_{i+1} \to U_{pi}/U_{pi+1}.$$

Proof. Choose a generator π for the prime ideal $\mathfrak{P}$. If $i > e_0$, so that $pi > i+e$, then the binomial expansion takes the form

$$(1+\pi^i\alpha)^p \equiv 1 + p\pi^i\alpha \mod U_{i+e+1}.$$

Hence the p-th power homomorphism induces an isomorphism

$$U_i/U_{i+1} \to U_{i+e}/U_{i+e+1}.$$

Using an easy successive approximation argument, it follows that each element of U_{i+e} has a unique p-th root in U_i.

REMARK A.5. Comparing the p-th power homomorphism $U_1 \to U_1$ with the corresponding homomorphism $U_1/U_i \to U_1/U_{i+e}$ for large i, we conclude that the cokernel $U_1/U_1{}^p$ has order either p^{ef+1} or p^{ef} according as F does or does not contain the p-th roots of unity. (Compare Lang, *Algebraic Number Theory*, p. 47.) If F has characteristic p, then this cokernel is infinite.

The proof of A.4 continues as follows. If $i < e_0$, then

$$(1+\pi^i\alpha)^p \equiv 1 + \pi^{pi}\alpha^p \mod U_{pi+1}.$$

Since the correspondence $\alpha \mapsto \alpha^p$ induces an automorphism of the residue class field, we obtain an induced isomorphism

$$U_i/U_{i+1} \to U_{pi}/U_{pi+1}.$$

Finally suppose that $i = e_0$. The kernel and cokernel of the p-th power homomorphism $U_{e_0} \to U_{e_0+e}$ can be identified with the kernel and cokernel of the induced homomorphism

$$U_{e_0}/U_{e_0+1} \to U_{e_0+e}/U_{e_0+e+1}.$$

But these are finite groups of the same order, so the kernel and cokernel have the same order. The kernel, consisting of all p-th roots of unity in U_{e_0}, certainly cannot have more than p elements.

To complete the proof, we must show that a primitive p-th root of unity, if it exists in F, must lie in U_{e_0}. But otherwise it would represent an element in the kernel of

$$U_i/U_{i+1} \to U_{pi}/U_{pi+1}$$

for some $i < e_0$, which is impossible. This proves A.4. ∎

It will be convenient to introduce an integer p^s, associated with any local field F of residue class field characteristic p, as follows.

DEFINITION. If F contains the p-th roots of unity, let p^s be the highest power of p dividing pe_0. If F contains only the trivial p-th root of unity, we set $p^s = 1$.

Our study of Steinberg symbols on F will be based on the following result, which will be proved presently. Choose some fixed prime element π. Let n be any positive integer.

MAIN LEMMA A.6. *Given any* $u_n \in U_n$, *there exists an element* u_{n+1} *in* U_{n+1} *satisfying the equation*

$$\{\pi, u_n\}^{p^s} = \{\pi, u_{n+1}\}^{p^s}.$$

Hence, starting with any u_1, we can inductively construct a sequence $u_1, u_2, u_3, \ldots$ with

$$\{\pi, u_1\}^{p^s} = \{\pi, u_2\}^{p^s} = \ldots = \{\pi, u_n\}^{p^s}.$$

In other words any u_1 can be approximated arbitrarily closely by an element of the form u_1/u_n which is "orthogonal" to π^{p^s} under the pairing $\{\ ,\ \}$.

Now let $c(\alpha,\beta)$ be a weakly continuous Steinberg symbol on F. Since u_1/u_n converges to u_1 as $n \to \infty$, and since

$$c(\pi^{p^s}, u_1/u_n) = 1,$$

it follows by weak continuity that

$$c(\pi, u_1)^{p^s} = c(\pi^{p^s}, u_1) = 1.$$

But any other prime element πu could be used in place of π throughout this argument. Since the multiplicative group $F^{\bullet}$ is generated by the set of all such prime elements, this proves the following.

COROLLARY A.7. *If* c *is weakly continuous, then* $c(\alpha, u_1)^{p^s} = 1$ *for all* α *in* $F^{\bullet}$ *and all* u_1 *in* U_1.

We can now give a good qualitative description of the symbol $c : F^\bullet \times F^\bullet \to A$. It will be convenient to introduce the notation $c(F^\bullet \otimes F^\bullet)$ for the subgroup of A generated by $c(\alpha,\beta)$ for all α and β in $F^\bullet$.

COROLLARY A.8. *If* c *is weakly continuous, then the image* $c(F^\bullet \otimes F^\bullet) \subset A$ *splits as the direct sum of a finite cyclic group* A_1 *of order dividing* $q-1$ *and a finite group* A_2 *of exponent** p^s. *There is one and only one homomorphism* $(\Lambda/\mathfrak{P})^\bullet \to A_1$ *which carries the tame symbol* $d_v(\alpha,\beta)$ *of* 11.5 *to the* A_1-*component of* $c(\alpha,\beta)$.

Proof of A.8 (still assuming A.6). Define an auxilliary Steinberg symbol d by

$$d(\alpha,\beta) = c(\alpha,\beta)^{p^s}.$$

For any u and w in $\Lambda^\bullet$, it follows from A.7 that the symbol $d(u,v)$ depends only on the images of u and v in the residue class field. But any Steinberg symbol on a finite field is trivial by 9.9, so $d(u,v) = 1$.

For any α and β in $F^\bullet$, setting $\alpha = \pi^i \eta u_1$ and $\beta = \pi^j \eta' u'_1$ (compare A.3), it follows that

$$d(\alpha,\beta) = d(\pi,\pi)^{ij} d(\eta,\pi)^j d(\pi,\eta')^i$$

is equal to

$$d((-1)^{ij}\eta^j/\eta'^i, \pi).$$

(Note that $(-1)^{ij}$ belongs either to μ_{q-1} or to U_1 according as q is odd or even.) Hence $d(\alpha,\beta)$ is the image of the tame symbol $d_v(\alpha,\beta)$ under the homomorphism

$$(\eta \bmod \mathfrak{P}) \mapsto d(\eta,\pi)$$

from $(\Lambda/\mathfrak{P})^\bullet$ to A. In particular it follows that the image $d(F^\bullet \otimes F^\bullet)$ is cyclic, of order dividing $q-1$.

Let $N = p^s(q-1)$. Then $c(\alpha,\beta)^N = d(\alpha,\beta)^{q-1} = 1$ for all α,β. Since the subgroup $(F^\bullet)^N \subset F^\bullet$ has finite index by A.3 and A.4, it follows that $c(F^\bullet \otimes F^\bullet)$ is a finite group, and the conclusions of A.8 follow easily. ∎

The proof of A.6 will be based on the following elementary remark.

* In other words $a^{p^s} = 1$ for all $a \in A_2$. For the definition of p^s see the discussion preceding A.6.

LEMMA A.9. *The equality* $\{\pi, (1-\pi^n\eta)^n\} = 1$ *holds for every* η *in* μ_{q-1} *and every* $n \geq 1$.

For the element $1 - \pi^n\eta$ has a (q−1)-st root w in U_n, hence

$$\{\eta, 1-\pi^n\eta\} = \{\eta, w^{q-1}\} = \{\eta^{q-1}, w\} = 1,$$

and it follows that

$$\{\pi, (1-\pi^n\eta)^n\} = \{\pi^n, 1-\pi^n\eta\} = \{\pi^n\eta, 1-\pi^n\eta\} = 1. \blacksquare$$

Proof of Lemma A.6. It will be convenient to work with the Steinberg symbol c defined by

$$c(\alpha,\beta) = \{\alpha,\beta\}^{p^s} \in K_2F.$$

We must show the following. *Every* u_n *in* U_n *is congruent modulo* U_{n+1} *to an element* γ *satisfying* $c(\pi,\gamma) = 1$. The proof, by induction on n, will be divided into four cases.

Case 1. If n is relatively prime to p, then the given element u_n has a unique n-th root w in U_n. We may assume that $w \notin U_{n+1}$. (For if $w \in U_{n+1}$ then $u_n \in U_{n+1}$, and we can simply choose $\gamma = 1$.) Writing $1 - w$ as a product

$$1 - w \in \pi^n\eta U_1$$

by A.3, it follows that

$$w \equiv 1 - \pi^n\eta \mod U_{n+1},$$

and hence

$$u_n = w^n \equiv (1-\pi^n\eta)^n \mod U_{n+1}.$$

Setting $\gamma = (1-\pi^n\eta)^n$, the equality $c(\pi,\gamma) = 1$ follows from A.9.

Case 2. If $n \equiv 0 \mod p$, and $n < pe_0$, then we can choose $w \in U_{n/p}$ so that

$$w^p \equiv u_n \mod U_{n+1}.$$

By the induction hypothesis, there exists an element $\gamma \equiv w \mod U_{(n/p)+1}$ satisfying $c(\pi,\gamma) = 1$. The element γ^p will then satisfy the required conditions $\gamma^p \equiv u_n \mod U_{n+1}$ and $c(\pi,\gamma^p) = 1$.

Case 3. If $n > pe_0$ the proof is the same except that one chooses $w \in U_{n-e}$ with $w^p = u_n$, and then chooses $\gamma \equiv w \mod U_{n-e+1}$ with $c(\pi,\gamma) = 1$. Again the p-th power $\gamma^p \equiv u_n \mod U_{n+1}$ satisfies the required equation $c(\pi,\gamma^p) = 1$.

Case 4. If $n = pe_0$, recall that the p-th power homomorphism

$$U_{e_0} \to U_n$$

has cokernel which is either trivial or cyclic of order p. If the cokernel is trivial, we can proceed just as in Cases 2 and 3 above. Suppose then that F contains the p-th roots of unity, so that the cokernel is non-trivial. Set $n = pe_0$ equal to $p^s n_1$ with n_1 relatively prime to p. Now choose a generator for the cokernel, and represent it by an element δ_π in U_n which has the form

$$\delta_\pi = (1-\pi^n\eta)^{n_1}$$

with $\eta \in \mu_{q-1}$. Then

$$c(\pi,\delta_\pi) = \{\pi,\delta_\pi\}^{p^s} = 1$$

by A.9.

Given u_n, we can now choose a power δ_π^i so that

$$u_n \equiv \delta_\pi^i \mod (U_{e_0})^p .$$

Setting $u_n = w^p\delta_\pi^i$ with $w \in U_{e_0}$, and setting

$$w \equiv \gamma \mod U_{e_0+1}$$

with $c(\pi,\gamma) = 1$, as in Cases 2 and 3, we obtain

$$u_n \equiv \gamma^p\delta_\pi^i \mod U_{n+1}$$

with $c(\pi,\gamma^p\delta_\pi^i) = 1$ as required. This completes the proof of Lemma A.6. ∎

In order to obtain a more precise description of weakly continuous Steinberg symbols, we will need several lemmas concerning the structure of $K_2F/(K_2F)^p$.

Suppose that F contains the p-th roots of unity, so that the p-th power homomorphism

$$U_{e_0} \to U_{pe_0}$$

has cokernel of order p.

DEFINITION. Any element δ of U_{pe_0} which does not have a p-th root in U_{e_0} (and hence does not have a p-th root in $F^\bullet$) will be called a *distinguished unit* of F.

LEMMA A.10. *If π is a prime element and δ a distinguished unit, then for any u_1 in U_1 the symbol $\{\pi,u_1\}$ is congruent to a power of $\{\pi,\delta\}$ modulo $(K_2F)^p$.*

Proof. Just as in the proof of A.6, starting with u_1 one can inductively construct a sequence $u_1, u_2, \ldots, u_{pe_0}$ so that

$$\{\pi,u_1\} = \{\pi,u_2\} = \ldots = \{\pi,u_{pe_0}\} .$$

But u_{pe_0} is congruent to a power of δ modulo $(F^\bullet)^p$, and the conclusion follows. ∎

LEMMA A.11. *If δ is a distinguished unit, then $\{u,\delta\} \equiv 1$ mod $(K_2F)^p$ for all $u \in \Lambda^\bullet$.*

Proof. (Compare O'Meara, *Introduction to Quadratic Forms*, p. 165.) We will show that the equation

$$\delta x^p + uy^p = 1$$

has a solution x,y in Λ. This will imply that

$$\{u,\delta\} = \{x,u\}^p \{\delta,y\}^p \{x,y\}^{p^2} \equiv 1$$

as required.

Setting $\delta = 1 - \pi^{pe_0}v$, choose $w \in \Lambda^\bullet$ so that

$$w^p \equiv v/u \mod \mathfrak{P} ,$$

and let $z = \pi^{e_0}w$. Then

$$\delta + uz^p \equiv 1 \mod \mathfrak{P}^{pe_0+1} .$$

Using A.4, it follows that the equation

$$(\delta+uz^p)\, x^p = 1$$

has a solution x. Now, setting $y = zx$, we obtain $\delta x^p + uy^p = 1$, which completes the proof. Combining these two lemmas we obtain:

COROLLARY A.12. *If* F *contains the* p-*th roots of unity, then* $K_2F/(K_2F)^p$ *is cyclic of order* p *with generator* $\{\pi,\delta\}$ *modulo* $(K_2F)^p$.

Proof. The existence of the p-th power norm residue symbol (§15.8) implies that $K_2F/(K_2F)^p$ is non-trivial. To show that it is cyclic, choose some prime element π, and consider an arbitrary generator $\{\alpha,\beta\}$ of K_2F. Setting $\beta = (-\pi)^i \eta u_1$ by A.3, the symbol

$$\{\pi,\beta\} = \{\pi,\eta u_1\} \equiv \{\pi,u_1\} \mod (K_2F)^p$$

is congruent to a power of $\{\pi,\delta\}$ by A.10. Let $\alpha = \pi^j u$. Using the prime πu in place of π in the argument above, we see that $\{\pi u,\beta\}$ is congruent to a power of $\{\pi u,\delta\}$. But $\{\pi u,\delta\} \equiv \{\pi,\delta\}$ by A.11, and the conclusion follows easily. ∎

Still assuming that F contains the p-th roots of unity, we will prove the following.

LEMMA A.13. *If* β *does not have a* p-*th root in* $F^\bullet$, *then there exists* α *in* $F^\bullet$ *so that* $\{\alpha,\beta\} \not\equiv 1 \mod (K_2F)^p$.

Proof. This statement follows immediately from standard properties of the p-th power norm residue symbol, as described in §15.10. (Compare Serre, *Corps Locaux*, p. 215.) A direct proof can be given as follows. (Compare O'Meara, p. 166.) Let $\beta = \pi^i u$ with $u \in \Lambda^\bullet$.

Case 1. If the value $v(\beta) = i$ is relatively prime to p, then

$$\{\delta,\beta\} = \{\delta,\pi^i\} \not\equiv 1$$

by A.11 and A.12.

Case 2. If $i \equiv 0 \mod p$, then $\beta \equiv u \mod (F^\bullet)^p$. In fact setting $u = \eta u_1$ we have

$$\beta \equiv u_1 \mod (F^\bullet)^p.$$

Let n *be the largest integer such that*

$$\beta \equiv u_n \mod (F^\bullet)^p$$

for some u_n *in* U_n. Such a largest integer must exist, and satisfy $n \le pe_0$; for if this congruence were true with $n > pe_0$ it would follow from A.4 that β had a p-th root; thus contradicting the hypothesis.

If $n = pe_0$, then u_n must be a distinguished unit. Choosing $\alpha = \pi$, it follows that

$$\{\pi,\beta\} \equiv \{\pi,u_n\} \not\equiv 1$$

as required.

Assume then that $0 < n < pe_0$. The integer n must be relatively prime to p, since otherwise by A.4 the element u_n would be congruent, modulo $(F^\bullet)^p$, to some u_{n+1}. Choose

$$\alpha = 1 - u_n/\delta.$$

Then $\{\alpha,u_n/\delta\} = 1$, and

$$v(\alpha) = n \not\equiv 0 \mod p,$$

so it follows that

$$\{\alpha,\beta\} \equiv \{\alpha,u_n\} = \{\alpha,\delta\} \equiv \{\pi^n,\delta\} \not\equiv 1,$$

as in Case 1. This proves Lemma A.13. ∎

Now let F be any local field, and let m be the number of roots of unity in F.

THEOREM A.14 (Moore). *The group* K_2F *is the direct sum of a cyclic group of order* m *and an infinitely divisible group* $(K_2F)^m$. *A Steinberg symbol* $c : F^\bullet \times F^\bullet \to A$ *is weakly continuous if and only if the associated homomorphism* $K_2F \to A$ *annihilates this subgroup* $(K_2F)^m$.

Thus if c is weakly continuous, then the image $c(F^\bullet \otimes F^\bullet)$ is a cyclic group of order dividing m; and conversely if $c(F^\bullet \otimes F^\bullet)$ is finite then c is weakly continuous.

The m-th power norm residue symbol $(\alpha,\beta)_F \in \mu_m$ evidently induces a surjection $K_2F/(K_2F)^m \to \mu_m$, so it follows that every weakly continuous symbol is a homomorphic image of $(\alpha,\beta)_F$.

Proof of A.14. Let $c(\alpha,\beta)$ be weakly continuous. Then according to A.8 the image $c(A \otimes A)$ splits as a direct sum $A_1 \oplus A_2$ where A_1 is cyclic of exponent $q-1$ and A_2 is finite of exponent p^s. This p-primary component A_2 actually occurs only if F contains the p-th roots of unity. But if F does contain the p-th roots of unity, then $K_2F/(K_2F)^p$ is cyclic by A.12. This clearly implies that $K_2F/(K_2F)^{p^s}$ is cyclic, and hence its homomorphic image A_2 must be cyclic. Thus the image $c(F^\bullet \otimes F^\bullet)$ is a finite cyclic group of order dividing $p^s(q-1)$.

To obtain a sharper upper bound for the order when $p^s \neq 1$, we proceed as follows. Let ξ be a primitive m-th root of unity. We are assuming that $p|m$, so ξ cannot have a p-th root in $F^\bullet$, and by A.13 there exists an element α_0 with

$$\{\alpha_0,\xi\} \not\equiv 1 \mod (K_2F)^p .$$

By A.12 this symbol $\{\alpha_0,\xi\}$ must generate K_2F modulo $(K_2F)^p$, and it follows immediately that $\{\alpha_0,\xi\}$ generates K_2F modulo $(K_2F)^{p^s}$. Hence the p-primary component of $c(\alpha_0,\xi)$ generates the group $A_2 \subset c(F^\bullet \otimes F^\bullet)$. Since $c(\alpha_0,\xi)^m = c(\alpha_0,\xi^m) = 1$, it follows that $(A_2)^m = 1$, and hence $c(\alpha,\beta)^m = 1$ for all α and β.

Proof that $(K_2F)^m$ *is an infinitely divisible group.* First let r be any prime number other than the characteristic of F. (In the characteristic zero case, r can be any prime number.) By A.3 and A.4 the subgroup $(F^\bullet)^{rm} \subset F^\bullet$ is open. So the symbol $\{\alpha,\beta\}$ modulo $(K_2F)^{rm}$ is locally constant as a function of α and β, and hence weakly continuous. The argument above then shows that

$$\{\alpha,\beta\}^m \equiv 1 \mod (K_2F)^{rm} .$$

Thus every generator $\{\alpha,\beta\}^m$ of $(K_2F)^m$ has an r-th root within $(K_2F)^m$.

Now suppose that F has characteristic p. Then we must also prove that every element of K_2F has a p-th root. The field F must be isomorphic to the field of formal power series in one variable π over a finite field. Consider the inseparable extension $E = F(\sqrt[p]{\pi})$ of degree p over F. Clearly every element of F has one and only one p-th root in E. In other words the norm homomorphism from $E^\bullet$ to $F^\bullet$, given by

$$\text{norm}(\varepsilon) = \varepsilon^p,$$

is bijective. Consider any generator $\{\alpha,\beta\}$ of K_2F. If β has a p-th root in F, then certainly $\{\alpha,\beta\} \equiv 1 \mod (K_2F)^p$. But otherwise the extension field $F(\sqrt[p]{\beta})$ can be identified with E. Hence α is a norm from $F(\sqrt[p]{\beta})$, and it follows from §14.3 that $\{\alpha,\beta\} \equiv 1 \mod (K_2F)^p$. Thus every generator of K_2F has a p-th root, and it follows that every element of $(K_2F)^m$ has a p-th root.

Thus $(K_2F)^m$ is divisible, and the exact sequence

$$1 \to (K_2F)^m \to K_2F \to K_2F/(K_2F)^m \to 1$$

evidently splits. The right hand group is cyclic, since the symbol $\{\alpha,\beta\}$ modulo $(K_2F)^m$ is weakly continuous; and using the m-th power norm residue symbol we see that its order is precisely m. This completes the proof. ∎

COROLLARY A.15 (Bass, Tate). *If* $E \supset F$ *are local fields, then the transfer homomorphism* $\iota^*: K_2E \to K_2F$ *is surjective.*

(By way of contrast, the transfer $K_2C \to K_2R$ clearly is not surjective.)

Proof. First suppose that the extension $E \supset F$ is completely non-abelian, in the sense that no intermediate field, except F itself, is Galois over F with abelian Galois group. Then it follows from local class field theory that the norm homomorphism $E^\bullet \to F^\bullet$ is surjective. (See for example *Corps Locaux*, p. 180 or Cassels and Fröhlich, p. 143.) Using the identity

$$\iota^*\{\varepsilon, \iota(\beta)\} = \{\text{norm}(\varepsilon),\beta\}$$

of §14, this implies that the homomorphism $\iota^*: K_2E \to K_2F$ is also surjective.

Next suppose that E is an abelian extension of F of prime degree r, where r divides the number of roots of unity in F. Then the quotient $F^\bullet/(F^\bullet)^r$ contains at least r^2 distinct elements. But according to local class field theory the quotient

$$F^\bullet/\text{norm } E^\bullet$$

is cyclic of order r. Hence there exists some $\text{norm}(\varepsilon)$ which has no r-th root in $F^\bullet$. Choose β so that

$$\{\mathrm{norm}(\varepsilon),\ \beta\} \not\equiv 1 \mod (K_2F)^r .$$

(If r is the residue class field characteristic, then this is possible by A.13, while otherwise the existence of such an element β follows easily from 11.5.) Now the identity $\iota^*\{\varepsilon, \iota(\beta)\} = \{\mathrm{norm}(\varepsilon), \beta\}$ shows that the transfer is surjective modulo $(K_2F)^r$. But the identity

$$\iota^*\{\iota(\alpha), \iota(\beta)\} = \{\mathrm{norm}\ \iota(\alpha), \beta\} = \{\alpha,\beta\}^r$$

shows that every element of $(K_2F)^r$ is in the image ι^*K_2E.

Finally, suppose that the degree d of E over F is relatively prime to the number of roots of unity in F. The argument above shows that every element of $(K_2F)^d$ is in ι^*K_2E. But $K_2F = (K_2F)^d$ by A.14.

Since every finite extension of F can be built up as a tower $F = F_0 \subset F_1 \subset \ldots \subset F_k = E$ where each $F_{i+1} \supset F_i$ fits into one of these three cases, this completes the proof. ∎

The Functor K_2^{top}

The preceding results can be expressed more concisely by introducing a group $K_2^{top}\Lambda$ associated with a topological ring Λ. Unfortunately it is not quite clear in what generality such a group can be defined.

Let G be a Hausdorff topological group. Recall (§11.4) that a central extension

$$1 \to C \xrightarrow{\iota} X \xrightarrow{\psi} G \to 1$$

is called *topological* if C and X are Hausdorff topological groups, ι is continuous and closed, and ψ is continuous and open. Such a topological central extension will be called *universal* if, for every other topological central extension $1 \to C' \to X' \to G \to 1$, there is one and only one continuous homomorphism from X to X′ over G. If a universal topological central extension exists, then its kernel C will be called Π_1G, and we will say briefly that Π_1G *is defined.* Evidently Π_1G, if defined, is a commutative Hausdorff topological group. (Compare Moore, Publ. Math. I.H.E.S. 35, p. 13. I do not know whether Moore's π_1 is the same as our Π_1.)

If Π_1G is defined, then an argument similar to that in §5.5 shows that the commutator subgroup [G,G] must be dense in G.

For any topological ring Λ we would like to define $K_2^{top}\Lambda$ as the direct limit, as $n \to \infty$, of the groups $\Pi_1 E(n,\Lambda)$. Here $E(n,\Lambda)$ is topologized as a subset of $GL(n,\Lambda)$, which is topologized as the set of all pairs A, B of $n \times n$ matrices satisfying $AB = BA = I$. Here is one case in which this definition makes sense.

LEMMA A.16. *If* Λ *is a commutative Hausdorff topological ring (i.e., with continuous addition and multiplication), and if the group* $\Lambda^{\bullet}$ *of units is an open subset of* Λ *with continuous division, then* $E(n,\Lambda)$ *possesses a universal topological central extension, for* $n \geq 3$.

Thus the topological group $\Pi_1 E(n,\Lambda)$ is defined, and the direct limit

$$K_2^{top}\Lambda = \varinjlim \Pi_1 E(n,\Lambda)$$

is defined. One hopes that $K_2^{top}\Lambda$, with the direct limit topology, is itself a Hausdorff topological group.

Proof of A.16. Proceeding as in §7.4, it is not difficult to construct a neighborhood N of I in $SL(n,\Lambda)$ so that every element A of N can be expressed as the product of $n^2 + 5n - 6$ elementary matrices, each of which depends continuously on A. Let $\hat{E}$ be the universal central extension of $E(n,\Lambda)$. Lifting each elementary matrix to $\hat{E}$, as in §5.10 or §11.4, we obtain a section

$$s : N \to \hat{E}.$$

Let $\mathcal{T}$ be the largest topology on $\hat{E}$ such that

(1) the section s is continuous

(2) the product operation $\hat{E} \times \hat{E} \to \hat{E}$ is continuous, and

(3) the symmetry $e \mapsto e^{-1}$ on $\hat{E}$ is continuous.

More precisely, as basis for $\mathcal{T}$ take all finite intersections $U_1 \cap \ldots \cap U_k$ where each U_i is open with respect to some topology satisfying (1), (2) and (3). Then it is easy to verify that $\mathcal{T}$ itself is a topology satisfying these three conditions.

Of course $\mathcal{T}$ will not, in general, satisfy the Hausdorff axiom. To correct this, we pass to the quotient group $X = \hat{E}/\text{Closure}(1)$. Then X is a Hausdorff topological group (compare Bourbaki, *General Topology*, III: *Topological Groups*, §2.6), and the surjection

$$X \to E(n,\Lambda)$$

is the required universal topological central extension. Further details will be left to the reader. ∎

In the case of a topological field F, the above definition of $K_2^{top}F$ acquires a little more substance. Let us say that a Steinberg symbol $c : F^\bullet \times F^\bullet \to A$ is *strictly continuous* if c is continuous as a function of two variables and satisfies Matsumoto's condition $\lim_{u,v\to 0} c(u,1+uv) = 1$, and if the target group A is a Hausdorff commutative topological group. (For the locally compact fields studied earlier, it is clear that weakly continuous implies strictly continuous, using the discrete topology on A.)

LEMMA A.17. *If* F *is a Hausdorff topological field (continuous addition, multiplication and division), then the topological groups* $\Pi_1 E(n,\Lambda)$ *are canonically isomorphic to each other (for* $n \geq 5$) *and to their direct limit* K_2^{top}. *The natural pairing*

$$u,v \mapsto \{u,v\}^{top}$$

from $F^\bullet \times F^\bullet$ *to* $K_2^{top}F$ *is strictly continuous; and any other strictly continuous Steinberg symbol* $c : F^\bullet \times F^\bullet \to A$ *is the image of* $\{u,v\}^{top}$ *under a continuous homomorphism* $K_2^{top}F \to A$.

Proof. This follows immediately from Matsumoto's Theorem 11.4. ∎

Thus if F is a local field or the field of real numbers, containing m roots of unity, we see that the m-th power norm residue symbol $(u,v)_F$ gives rise to an isomorphism

$$K_2^{top}F \overset{\cong}{\to} \mu_m .$$

(Compare A.1 and A.14.) On the other hand, for the field C of complex numbers, we see that $K_2^{top}C$ is trivial.

These statements can be extended slightly by means of the following.

LEMMA A.18. *Suppose that* F_0 *is a dense subfield of the topological field* F. *Then* $K_2^{top}F_0$ *is a dense subgroup of* $K_2^{top}F$.

Now if F is locally compact, then $K_2^{top}F$ is discrete, and it follows that $K_2^{top}F_0 = K_2^{top}F$.

Proof. According to Bourbaki, *General Topology* III: *Topological Groups*, §6.5, the pairing $u,v \mapsto \{u,v\}^{top}$ from $F_0^\bullet \times F_0^\bullet$ to $K_2^{top}F_0$ extends to a continuous bilinear pairing

$$c : F^\bullet \times F^\bullet \to A$$

where A denotes the "completion" of the topological group $K_2^{top}F_0$. Since a Hausdorff topological group is necessarily regular, it is easy to verify that c is a strictly continuous Steinberg symbol. Hence c gives rise to a continuous homomorphism $c' : K_2^{top}F \to A$. Now extend c' to the completion B of $K_2^{top}F$. Since the composition $A \to B \to A$ is the identity, and since the image of A is clearly dense in B, it follows that $A \cong B$, which completes the proof. ■

INDEX

GPSR Authorized Representative: Easy Access System Europe - Mustamäe tee 50, 10621 Tallinn, Estonia, gpsr.requests@easproject.com

www.ingramcontent.com/pod-product-compliance
Ingram Content Group UK Ltd.
Pitfield, Milton Keynes, MK11 3LW, UK
UKHW011524220525
458817UK00001B/63

* 9 7 8 0 6 9 1 0 8 1 0 1 4 *